D20
BONUS
DP

Synoptic climatology in environmental analysis

STUDIES IN CLIMATOLOGY SERIES
Edited by Professor S. Gregory, University of Sheffield

A. M. Carleton, *Satellite Remote Sensing in Climatology*

C. M. Goodess, J. P. Palutikof and T. D. Davies, *The Nature and Causes of Climate Change*

B. Yarnal, *Synoptic Climatology in Environmental Analysis*

Other titles are in preparation

Synoptic climatology in environmental analysis

A primer

BRENT YARNAL

Belhaven Press
London and Florida

Belhaven Press
(a division of Pinter Publishers)
25 Floral Street, Covent Garden, London, WC2E 9DS, United
Kingdom

First published in 1993

British Library Cataloguing in Publication Data
A CIP catalogue record for this book is available from
the British Library.

ISBN 1 85293 117 5

Distributed in North America by
CRC Press Inc., 2000 Corporate Blvd., N.W.,
Boca Raton, Florida, 33431

Library of Congress Cataloging-in-Publication Data
A CIP catalog record for this book is available from the Library of
Congress.

Typeset by The Castlefield Press Ltd., Wellingborough, Northants
Printed and bound in Great Britain by Biddles Ltd., Guildford and
King's Lynn

Contents

List of figures

List of figures

List of tables

Foreword

This book is a primer in synoptic climatology. It aims to teach the reader what synoptic climatology is and how to apply it to environmental problems. It takes the perspective of the geographical climatologist because most research in synoptic climatology uses this empirical, deductive approach. The book is written in qualitative rather than quantitative terms. Only a few fairly basic equations are included, although references to more quantitative assessments are provided where appropriate.

The structure of the book is simple. Chapter 1 supplies the conceptual background needed to understand and use synoptic climatology. Chapters 2–5 form the heart of the book, presenting the main classification methods used by synoptic climatologists. In each chapter, I explain how the technique works, and then I review the research of other investigators who have used it. This gives the reader an idea of the environmental problems to which the technique can be applied. Step-by-step worked examples are a special feature of these four chapters. In Chapter 6 the performance of each of these worked examples is assessed. I also present a series of environmental scenarios to illustrate how synoptic climatology relates the atmospheric circulation to the surface environment. The scenarios include urban air quality, acid rain, agriculture, and fluvial hydrology. The book concludes with a call for increased collaboration between synoptic climatologists and scientists from other fields.

Synoptic climatology has much to offer scientists from many fields. Throughout this book I try to demonstrate the value of synoptic climatology in environmental analysis; my objective is to make synoptic climatology easy to understand, thus clarifying it for its specialists, students, and prospective collaborators.

Editor's preface

From being a somewhat disregarded minority interest within the field of meteorology, or equally a minority interest in associated fields such as geography, climatology has become — within the past ten to fifteen years — a central focus of innumerable disciplines concerned with the present and future environment of this planet. No longer seen as being concerned solely with average weather, or with general descriptions of recent conditions, it spans time-scales from the historical and geological past to probable events in the future, requires massive computing facilities for its statistical and mathematical modelling approaches, draws on the information from and technology of space satellites, and focuses on problems of prime importance to the future well-being of mankind.

Reflecting this growth in scale, quantity, scientific depth and applied relevance of research in climatology has been an equivalent growth in publications. The number of international journals specifically concerned with research papers in this field has increased considerably; conferences on climatic issues have proliferated, organized by a range of official or semi-official bodies, with volumes of proceedings or overall reports tending to be published for each of them; and various series of books concerned with specific aspects of the overall climatological field have appeared.

It is therefore reasonable to ask why another such series should be created, or at least where it fits into the overall picture of contemporary climatology. This series is primarily concerned with global climates of the present, how they have fluctuated or changed over the recent past, and the impact of such conditions on human society. This necessarily includes considerations of data sources, methodologies and theories as well as reviews and discussions of the evidence for the climatic conditions themselves. Such volumes will thus help to delimit and define the framework in the context of which the results of modelling studies, estimates of future climatic conditions, and assessments of environmental change need to be evaluated. This series comprises monographs or reviews of the current state of knowledge and understanding, essential for the researcher and perhaps also for the final year specialist student.

This third volume in the series admirably meets the aims and objectives outlined above. For approximately the last half-century, synoptic climatological studies have been published, their number and complexity increasing markedly over this period. For much of this time such enquiries have been linked to expanding and illuminating regional climatic accounts, and thus have had their *raison d'être* strictly in the meteorological/climatological field. The problem of the definition and specification of relevant weather types or synoptic patterns has bulked large in such enquiries, especially as the potentialities of computer analysis have been realized. The examination, review and evaluation of these various methodologies form the central core of this volume.

The author's contribution does not end there, however; he is especially concerned to illustrate and justify the role that synoptic climatology has played, and can increasingly play in the future, in the study and elucidation of a wide range of environmental problems. In addition, he also wants researchers utilizing a synoptic climatological approach to environmental issues to be well prepared to understand the range of possibilities that the methods make available. To this end, the presentation is essentially a personal one with the express purpose of providing a 'primer' for aspiring synoptic climatologists. Thus, not only does this volume provide an up-to-date review and assessment of the field of study, it also presents a mode of study and enquiry that can be followed by others.

S. Gregory
Sheffield
June 1992

Preface and acknowledgments

The coincidence of two events resulted in this book. First, Penn State's Department of Geography admitted a cohort of graduate students who wanted to learn synoptic-climatological theory and method. At the same time, I was struggling with the problem of which synoptic-climatological procedure was the best to apply to an environmental problem I was funded to study. To tackle both of these, I devised a year-long practicum in which I worked closely with five graduate students.

The practicum had several educational tasks. The investigative team:

— comprehensively reviewed the last two decades of synoptic climatology literature;
— worked out the procedures of each common synoptic-classification scheme;
— gained hands-on experience using each of the most popular classification techniques by devising a classification centered on western and central Pennsylvania;
— learned how to gather and prepare surface data for synoptic-climatological analysis by preparing environmental scenarios of urban air quality, acid rain, agriculture and fluvial hydrology;
— computed statistics comparing the relative performance of each classification for their scenario;
— mastered the mechanics of relating the synoptic classifications to the surface environment by developing full synoptic climatologies of each scenario.

Each investigator developed a full synoptic climatology for one of the environmental scenarios. In this way he acquired an appreciation of how to use synoptic climatology to analyze a specific environmental problem. Simultaneously, by following the progress of the other team members, he observed other problems and solutions. In sum, by experiencing the strengths and weaknesses of this scholarly process, the investigators came to understand how synoptic climatology works, as well as when it does not work.

As the organizer of this educational experience, I provided guidance to the investigative team. Meanwhile, from my omnipotent position I was able to study somewhat neutrally the processes of classifying the circulation data, developing scanarios, and relating the circulation to the environment. While being mindful of the calendar, I permitted the investigators to develop lines of thought that paralleled previous research, to start down blind alleys entered by other investigators, and, most important, to try out new ideas. I made allowance for uncertainties in scientific knowledge and for possible surprises. From this year-long program, the investigative team members became knowledgeable, experienced synoptic climatologists, while I was able to assemble and digest the information that forms the foundation of this book.

The other goal of the practicum was to determine which synoptic classification scheme was best for a particular environmental problem on which I was working. To be honest, at the beginning of the project, I had no idea which methodology was most suited to our problem. I did not realize that answering that seemingly simple question would lead me on an odyssey that has taken three years to complete.

To make sense of the jumble that synoptic climatology has become, I took it apart bit by bit and put it back together again. The form that the reassembled pieces took is this book. I think this approach would please my friends and colleagues from the post-modern school of analysis. Nevertheless, I owe much of my intellectual stance to the realists. Realists attempt to assess theory and method objectively and to seek the middle ground between opposing paradigms. They try to adopt what works, discard what does not work, and to promote compromise in theory and practice. From them I was able to develop a frank appraisal of the nature of synoptic climatology and the position from which I would develop this book. Let me explain.

Climatologists come from one of two philosophical directions. One uses inductive methods. This is the modern meteorological perspective, where all relations are reduced to mathematical formulae. Although meteorologists admit that nature is chaotic, they attempt to draw an envelope around climate systems (that is, make them closed systems) and to model the state variables explicitly. When life is too complicated to model, they reluctantly parameterize it. Observations of nature are secondary and often are considered as a step in the modeling process. The other perspective of climatology is more akin to the life and social sciences; these climatologists often come from the discipline of geography. To them, the natural world is infinitely complex, with varying processes operating at varying scales. They see the environment as an open system where external inputs are continually changing the relationships among variables. These climatologists admit that nature is messy and find it appropriate to explain the physics of the climate system through observation, statistical analysis, and deduction.

Synoptic climatologists favor the latter approach. Thus, in this book I acknowledge synoptic climatology's geographical heritage and focus on it. I do not deal with the minor number of studies that have approached synoptic climatological problems from the inductive, modeling perspective. None the less, in the concluding chapter, I make a pitch for synoptic climatologists to maintain their empirical, deductive roots but to use the more detailed, systematic approach of modern meteorology. There is little doubt in my mind that synoptic climatology could benefit from more scientific discipline.

Roger Barry and Allen Perry (1973) exhaustively surveyed the literature of synoptic climatology up to that time. They also presented the history of synoptic climatology. I see no reason to repeat this material, so I concentrate on the synoptic-climatological research of the last two decades. This modern period is quite distinct from the era covered by Barry and Perry, anyway. In

1974, the National Center for Atmospheric Research made available computer tapes containing long-term gridded pressure data. Coupled with ever-growing computer power, this event transformed the practice of synoptic climatology.

Unlike Barry and Perry, I do not attempt to present a synopsis of the physical processes that form the backbone of synoptic climatology. I assume that readers know the basics of climatology and meteorology, and especially the fundamentals of the atmospheric circulation. Those who are not familiar with these subjects, or those who wish to refresh their memories, should consult two recent publications. Jay Harman (1991) provides a lucid non-mathematical treatment of synoptic and dynamic processes in the westerlies. I have been using an earlier version of this monograph for years to teach beginning climate dynamics to undergraduates in both meteorology and geography. For a quantitative approach to mid-latitude weather systems, see Toby Carlson (1991), my colleague at Penn State. His conceptual and quantitative models are excellent.

Another subject area I do not cover is satellite-based synoptic climatology. Andrew Carleton covers this subject extensively in *Satellite Remote Sensing in Climatology* (Carleton, 1992). Again, I see little reason to duplicate material covered by another author.

Because this book is a primer, I adopt an active voice in my writing. Also, I treat the authors of studies as dynamic participants in the science of synoptic climatology, and I use direct, rather than indirect, references to them in most cases. I refer to authors as 'investigators' and talk about my students as the 'investigative team'. I participate actively by using the first person.

This book would not have been possible without the five graduate students who formed my investigative team. All have left Penn State and are pursuing promising careers. They are:

— Andrew Comrie, Assistant Professor, Department of Geography, University of Arizona;
— Maxx Dilley, AAAS Congressional Fellow, United States Agency for International Development;
— John Draves, Scientist, Hughes STX, Goddard Space Flight Center;
— Bruce Hewitson, Lecturer, Department of Geography and Environmental Sciences, University of Capetown;
— Ken Yelsey, State Geohydrologist, Vermont.

Special thanks must go to Andrew, who played Jeeves to my Bertie Wooster.

Many others deserve recognition for their part in helping to put together this book or stimulating me to action. The graphics are from the Deasy GeoGraphics Laboratory, Department of Geography, Penn State. Matt Tharp, Chief Cartographer, supervised the production of the graphics, which were drafted by Doug Smith, Mary Carpenter, David Wagner, and Jeremy Frick. John Skelly and Mike Simini, from Penn State's Department of Plant Pathology, were fellow recipients of the College of Agriculture grant that paid for this research. They saw the value of synoptic climatology from the beginning and came to me for help. Jim Lynch, School of Forest Resources at Penn State, provided me with the acid-rain, stream-flow, and stream-sulfate data. Eric Barron, Director of Penn State's Earth System Science Center (ESSC), arranged for me to get time off to complete the writing of this book. NASA's Earth Observing System program paid for that time through ESSC's "The Global Water Cycle: Extension Across the Earth Sciences" project — Eric Barron, Principal Investigator. This research was entirely supported by the Intercollege Research Program, College of Agriculture, Penn State: "The Climatology of Ozone as it Relates to Forest Health in Pennsylvania," — John Skelly, Mike Simini, and Brent Yarnal, Co-Principal Investigators.

Perhaps the nicest part of this project was writing the book. To get away from the hubbub of Penn State, my family retreated to North Carolina's Cape Hatteras for a semester. With their patience and understanding, I was able to think and write without the constant distractions that go along with faculty life. We all benefited, too, from the isolation, the sometimes wild winter weather, and the togetherness it brought to our little band. The last thanks, therefore, must go to Careen, Alison and Neil.

Brent Yarnal
Avon, North Carolina
April 1991

1 Introduction to synoptic climatology

Processes associated with many environmental problems such as air quality, acid rain, water quantity and water quality are strongly influenced by the circulation of the atmosphere. Surface ozone levels rise on sunny, warm days but fall dramatically when a rainstorm passes overhead. Rainfall in the northeastern United States is extremely acidic when preceded by several days of muggy southwesterly flow. Water levels go down and the concentrations of pollutants in that water go up when the storm track moves away from its normal position and drought prevails over an area. Scientists, environmental planners, and politicians can develop better strategies to mitigate the effects of such environmental problems if they understand how the atmospheric circulation controls environmental behavior.

By definition, the field of synoptic climatology studies the relationships between the atmospheric circulation and the surface environment of a region. Because synoptic climatology seeks to explain key interactions between the atmosphere and surface environment, it has great potential for basic and applied research in the environmental sciences. It also can contribute to the other fields of the atmospheric sciences and to geography.

Synoptic climatology is important to other atmospheric sciences because it synthesizes several fields of climatology.

— Dynamic climatology investigates the processes and patterns of air in motion; synoptic climatology focuses on the circulation. Therefore, investigators must incorporate the concepts of atmospheric dynamics in their research.
— One of the prime concerns of synoptic climatology is how variations in the atmospheric circulation induce changes in the surface environment. Therefore, the field of climatic change and variability is a necessary component of synoptic climatology.
— Regional climatology determines the similarity of climate within discrete areas, probing the causes and effects of related, parallel climatic processes within these regions. Synoptic climatology relates the circulation to a surface region; thus, regional climatology is integral to synoptic climatology.
— Physical climatology examines physical mass and energy fluxes in the environment, with primary emphasis on the earth's boundary layer. Synoptic climatology associates the circulation with these surface fluxes, and so synoptic climatologists must be familiar with physical processes.
— Finally, synoptic climatology is a form of applied climatology. The synoptic climatologist is primarily motivated by considerations of how variations in the atmospheric circulation affect the earth's surface where society operates.

Synoptic climatology also has special appeal to geographers, and, as a result, most synoptic climatologists are located in geography departments. From the above it is apparent that synoptic climatology involves the concepts of space, time, region, and surface environment that form the core of geographic inquiry. Furthermore, geographers stress a synthetic, holistic approach to science. Synoptic climatology's blend of diverse environmental and climatological fields, therefore, strikes a sympathetic chord with geographers.

Synoptic climatology provides investigators in the atmospheric, environmental, and geographical sciences with useful tools. For instance, dynamic and physical modelers in meteorology usually use average climatic conditions or individual cases to drive their models. Growing numbers of meteorologists are turning to synoptic climatology to determine the primary modes of atmospheric circulation, thus developing more realistic and representative scenarios for their model runs. In another example of synoptic climatology's usefulness, large field experiments in the atmospheric and environmental sciences cost millions of dollars to conduct. Team activities are specified well in advance; the appropriateness of that day's weather to the experiment is left to chance. Hidy (1988) stresses that synoptic climatology should be used to identify key meteorological configurations. Activating the team only when forecasters predict a suitable weather pattern saves time, effort and money; it also produces better results.

Besides being useful to others, synoptic climatology has its own intrinsic merits as a science. This book amply demonstrates that synoptic climatology conveys vast amounts of integrated knowledge about atmospheric dynamics, physical climatology, regional climates, and climatic variability and change. Although science holds great hopes for simulation models of atmosphere-surface interactions, at present synoptic climatology is "the only game in town." Synoptic climatology provides the immediate means for understanding how environmental processes and the atmospheric circulation are linked. With increasing incorporation of dynamic processes into synoptic-climatological analysis, synoptic climatology presents great promise as a bridge to functional simulations of the complex climate system.

This book is a guide to synoptic climatology. In it, I aim not only to describe what synoptic climatology is, what it has done, and what it can do, but also to teach the reader how to do it. I give practical schooling on the art and science of synoptic climatology by clearly laying out the methods of synoptic climatology, presenting numerous worked examples of synoptic classification, and developing scaled-down synoptic climatologies based on those worked examples. Before I can start instruction, however, I must develop a background on which to develop this work. The remainder of Chapter 1 provides the historical, theoretical, and structural environment that forms the basis of synoptic climatology.

Synoptic climatology in the last two decades

R.G. Barry and A.H. Perry (1973) laid down the history and basic principles of the field in their book, *Synoptic Climatology*. However, soon after the publication of this work, the computer revolution transformed synoptic climatology. This subject's reliance on large data sets and statistical analysis spurred rapid adoption of computer technology. A generation of synoptic climatologists trained on the computer have matured since the publication of *Synoptic Climatology*.

Paralleling the computer revolution, synoptic climatology has been on the ascendancy from the late 1970s, and a strong case can be made that it is now the dominant field in geographical climatology. For instance, many PhD-granting geography departments in the United States have at least one person who calls him- or herself a synoptic climatologist or someone who regularly conducts synoptic-climatological research. Most of these faculty members are under age 45. There is a practical, historical reason for this. Synoptic climatological research is difficult without the aid of a computer. Computer-literate climatologists were rare until the generation of scientists trained in the mid- to late 1970s entered the faculty labor force. These climatologists preferentially worked on synoptic problems, perhaps seeing the empty research niche and realizing the potential for significant contribution. At about this time, many senior climatologists, who learned their climatology during World War II and began their faculty work shortly thereafter, reached retirement age. Their positions were often filled by synoptic climatologists, partly because there were so many available and partly because their computer literacy put them at the leading edge of research. This caught the eye of faculty-recruiting committees. The fact that many synoptic climatologists were hired in the late 1970s and early 1980s also suggests another reason why this field is continuing to gain strength in American geographical climatology. This cohort is reaching the point where its large numbers, combined with its maturation, places it in a commanding position within geographical climatology.

The number of publications in synoptic climatology grew throughout the late 1970s and early 1980s, but Smithson (1988) noted a decline in publications starting in the mid-1980s. Harman and Winkler (1991) suggest that this reduction was not real, but merely fewer papers in the field have the words "synoptic climatology" in the title or key-words list. This is especially true if the articles appear in non-geographical or non-meteorological journals. Harman and Winkler, in fact, see an upturn in activity in synoptic climatology, while my literature review for this book suggests that more synoptic climatologists are publishing outside the mainstream geography and climatology press.

Theory

In this section, I introduce the theoretical framework needed to understand modern synoptic climatology and, in a more practical vein, the book rationale and structure. First I provide a working definition for the field which is used throughout the book. Then I present the different approaches to synoptic-climatological classification; I base much of the book's structure on the categories furnished here. Next, I address the goals of synoptic climatology, and then I list some of the key theoretical questions of the field. Following that, I describe the assumptions that all synoptic climatologists use. I conclude by revealing some of the theoretical problems that frustrate all synoptic climatologists.

Definition of synoptic climatology

In an early article, Court (1957) presented definitions of dynamic and synoptic climatology. Although these ultimately were not accepted, he did point out key distinctions between the fields that have stood the test of time. The first relates to scale. In the 1950s, dynamic climatologists tended to work at the large synoptic to planetary scales (say, greater than a few thousand kilometers in the horizontal dimension), while synoptic climatologists concentrated on smaller synoptic scales. This difference has become less pronounced over time. With the growth of computer power and computer-based data sets, today's synoptic climatologist often produces planetary-scale studies. None the less, the majority of modern synoptic climatologists favor the synoptic scale. Second, in Court's day, dynamic climatologists often employed variables such as velocity potential, momentum, energy, convergence, divergence and vorticity; synoptic climatologists used concepts like pressure pattern, air flow, air mass and map pattern. Few modern-day synoptic climatologists lack a foundation in atmospheric dynamics. However, the majority of synoptic climatologies still utilize the latter set of terms and variables. The third point to draw from Court's analysis is that synoptic climatology specifically relates the atmospheric circulation to surface climate, while dynamic climatology seldom concerns itself with surface connections to the free-air circulation.

In a commentary on Court's article, Lydolph (1957) stated that dynamic climatology "attempts a complete explanation of the states of the atmosphere." In contrast, synoptic climatology "attempts an explanation of local weather" by "using the known states of the atmosphere." He believed that climatologists study a continuum, extending from the driving thermodynamic forces to the resulting surface climate. Tongue in cheek, he called dynamic climatology "research-meteorologist climatology" and synoptic climatology "physical geographer-forecasting meteorologist climatology." More seriously, he proposed the terms *inductive climatology* for the former and *deductive climatology* for the latter.

The discussions of Court and Lydolph provided a sense of the difference between the two fields. Dynamic climatology takes a pure-science approach, but synoptic climatology sets applied-science objectives; synoptic climatology takes a deductive approach to science, while dynamic climatology uses induction. Still, there are numerous exceptions to these generalizations. The border between synoptic and dynamic climatology is a gray area, rather than a neatly defined black line. Furthermore, the two fields complement one another. Synoptic climatology gives purpose to dynamic climatology, and dynamic climatology provides a physical basis for synoptic-climatological relationships.

Following this dialogue, Barry and Perry (1973) set the course for most synoptic climatologies of the last two decades. They defined synoptic climatology as the study of the relationship between the atmospheric circulation and local or regional climates. They concluded that every synoptic climatology has two stages: (1) the classification of the atmospheric circulation; and (2) the assessment of the relationship between those categories and the region's *weather elements* (my italics).

Although the first of these stages has held true, the second has drifted somewhat in the intervening years. In the early 1970s, most synoptic classifications were related to weather elements. Since then, science has collected a wide array of non-meteorological environmental data. Furthermore, all climatologists now recognize that the climate system links atmospheric and non-atmospheric variables (for example, Yarnal *et al.*, 1987). As I will demonstrate through the literature review and environmental scenarios of this book, today's synoptic climatologist often works with non-meteorological but climatically-related variables such as surface ozone, crop yields, and water quantity and quality. Thus a more general but more accurate working definition states that *synoptic climatology relates the atmospheric circulation to the surface environment.* "Surface" in this case refers to the planetary boundary layer, which has an average thickness of about one kilometer. Throughout the book, I will use this simple definition to determine whether a study is a synoptic climatology or not.

Synoptic climatology, however, is more than just the cold relationship between two numbers, one from the atmosphere and one from the surface. These two numbers truly represent synoptic climatology: one embodies the state of the atmospheric circulation and the other incarnates the whole of the surface environment. The definition of *synoptic* most often associated with synoptic climatology is *simultaneous*, such as the simultaneous measurement of weather variables at the international synoptic hours (for example, 0000 GMT). However, the other definition of synoptic also applies; that is, *holistic*, or *all encompassing*. Putting together these two definitions of synoptic, synoptic climatology integrates the simultaneous atmospheric dynamics and coupled response of the surface environment. It includes all variables, scales, feedbacks, and non-linearities that are involved in that complex, coupled system.

There are two ways that scientists model such systems. One uses the inductive approach of the dynamic climatologist. The evolution of science and technology now means that this approach relies on mathematical computer models to link atmospheric motions and surface responses (Rayner, 1984). Such models certainly meet the definition of synoptic climatology; that is the models relate the circulation to the surface. This approach, however, does not provide a holistic, synoptic appraisal of the climate system. Even the most intricate mathematical model is a crude abstraction of reality. The second approach to modeling complex climatic systems utilizes the deductive approach. This strategy relies on statistics to link the circulation and the surface environment. A large proportion of today's climatologists finds this "black-box" approach to science unsatisfactory. At the same time, the opposite camp finds the highly abstracted mathematical models too unrealistic for their tastes.

This inductive–deductive split in science is not new and is certainly not unique to dynamic and synoptic climatology. As suggested earlier, I believe the two approaches are not mutually exclusive but complementary. The synoptic climatologist teases empirical relationships out of the observed record, providing the foundation upon which the dynamic modeler bases his or her work. The dynamicist, in turn, develops the theory which synopticians use to figure out those processes for which they should be looking. Despite the occasional acrimony, neither approach to large-scale climatology can exist without the other. In this book, I assume that synoptic climatology involves deductive, empirical research. The weight of the literature supports that assumption.

The polarization between these two camps is not perfect. Many empiricists observe the atmospheric circulation but do not relate their observations to the surface. To limit the scope of this book and keep the literature manageable, I treat these studies as dynamic climatologies. Few modelers attempt to link the circulation and the surface on more than a case-study basis. If they did expand their work to include repetitive, climatic perspectives, I would still probably classify them as dynamicists. The line between synoptic and dynamic climatology will become increasingly difficult to draw, however, as theory, modeling skills, and computer power all grow in the future. I return to this point in the concluding chapter.

Methodological approaches to synoptic climatology

All synoptic-climatological studies have four factors in common. First, they classify the atmospheric circulation in some way. In classification, individuals are grouped either according to similarities of properties or by relationships between objects (Semple and Green, 1984). In this case, "similarities of properties" might mean a combination of weather variables characteristic of an air mass, while "relationships between objects" could refer to pressure centers on a map surface. There are many important reasons for climatic classification (Balling, 1984). Among these, two are crucial to synoptic climatology: (1) classification schemes bring order, structure, and simplicity to the complex climate system (Harvey, 1969); and (2) classification provides an intellectual shorthand that conveys huge amounts of information via a simple label (Abler *et al.*, 1971). Thus, classification enables the synoptic climatologist to identify the essentials of the atmospheric circulation and to communicate this essence by means of an understandable, descriptive class name.

Second, all synoptic climatologies link at least two scales of analysis: the large-scale atmospheric circulation and the smaller-scale surface environment. Relating the physical processes active at varying scales is one of the fundamental issues facing earth and environmental scientists today (Rosswall *et al.*, 1988; Turner *et al.*, 1989). Synoptic climatologists have been addressing the problem of scale linkage for decades, using statistical rather than mathematical models. Indeed, Harman and Winkler (1991) believe that the cascade of linkages from larger to smaller scales and its ultimate relationship to the surface environment lies at the core of synoptic climatology.

Third, all synoptic climatologies study the effect of climatic variability on the surface environment. Within-season and interannual fluctuations in the atmospheric circulation determine the surface climate of a given period. One of the most important aspects of synoptic climatology is identifying the control that these variations in the circulation have on the environment.

The fourth commonality among synoptic climatologies is the focus on region as the fundamental spatial entity of concern. The scale of the atmospheric circulation under consideration may vary from planetary to small synoptic scale, but the circulation is always linked to the surface environment of a region. Although practicalities may restrict the analysis to surface data collected at one point, investigators assume those data represent the region.

From the shared starting point provided by these factors, approaches to synoptic climatology start to diverge. All synoptic-climatological studies use one of two fundamental approaches to classification. I will call these *circulation-to-environment* and *environment-to-circulation*, for lack of better descriptive terms. The distinction between the two approaches is in the way the atmospheric classification and the surface environment relate to one another. In the former case, the environmental data are assessed relative to the synoptic classes; in the latter, the synoptic classes fit criteria based on the environmental variable (Figure 1.1).

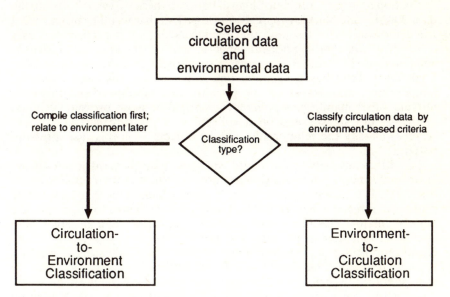

Figure 1.1 The two fundamental approaches to synoptic classification.

The investigator designs the synoptic classification to relate to a region in the circulation-to-environment approach. The classification is fairly general, typically representing the complete range of the atmospheric circulation over the area and the entire period for which data are available. Although the synoptic data might be chosen because they are believed to have a distinctive effect on a certain environmental variable, no environment-specific criteria are set for the circulation data's inclusion in the classification scheme. Simply, the classification of circulation data is independent of the environmental response. To illustrate this approach: say the investigator wishes to study the relationship between surface ozone and the atmospheric circulation. If he or she *first* compiles a classification of weather types over the region and *then* relates the classes to days with elevated ozone concentrations, the investigator is employing a circulation-to-environment scheme.

In contrast, the environment-to-circulation approach finds the investigator identifying the atmospheric circulation associated with specific environmental conditions at the surface. With this approach, the investigator sets environment-specific criteria for inclusion of the atmospheric circulation data. The synoptic classes are not independent of the environmental response. For example, in the hypothetical research on surface ozone, if the investigator only classifies the circulation when ozone concentrations are greater than 120 parts per billion (ppb), then the surface environment is dictating the selection of synoptic data.

As noted earlier, classifications group data according to similarities of properties or relationships between objects. Synoptic classifications, whether they use the circulation-to-environment or environment-to-circulation approach, often distinguish between these two ways to group the data. On the one hand, *synoptic types* classify similar weather properties. I will demonstrate, for instance, that eigenvector-based synoptic typing uses a blend of principal components analysis and cluster analysis to identify combinations of weather elements characteristic of different air masses. Each eigenvector-based synoptic type represents a distinct mixture of temperature, humidity, sky cover, pressure, and wind speed and direction. On the other hand, *map-pattern classifications* show pressure centers, gradients and signs found on a collection of pressure surfaces. These schemes classify the relationships among objects — in this case, pressure patterns.

The difference between synoptic types and map patterns is not always clear, and synoptic climatologists are not consistent in their application of the terms. The most important example of this occurs in the manual synoptic types presented in Chapter 2. The Lamb weather types, for instance, are based on manual classification of weather maps, so it seems logical that these should be called map patterns and not synoptic types. As part of the classification process, however, Lamb studies other indicators of weather besides pressure fields such as cloud patterns and the location of precipitation. In fact, in all manual schemes based on weather maps, investigators use a holistic combination of all weather elements, in addition to the pressure patterns, to determine a map's ultimate classification.

Consequently, these are synoptic types and not map patterns. I try to be consistent in my use of "synoptic types" and "map patterns" throughout this book.

Another major split in synoptic climatology cleaves the *manual* and *automated* classifications. In manual classification, each piece of data is handled and assigned to a synoptic class by the investigator, using a classification developed by another scientist or developing a new one. Automated procedures, also referred to as *computer-based* or *computer-assisted* methods, read digital data. Computers determine the classes and separate the data into those categories. Not all methods are purely manual or automated. For instance, compositing and indexing are computer assisted but rely on the investigator to determine the classification. Truthfully, no synoptic climatologist works without the benefit of a computer today. Although the classification technique might be manual, the investigator stores the classes in a spreadsheet and analyzes the data with a computer.

From the above and a review of the literature, I identify several classification methods used in modern synoptic climatology:

— Manual synoptic types;
— Correlation-based map patterns;
— Eigenvector-based synoptic types;
— Eigenvector-based map patterns;
— Eigenvector-based regionalizations;
— Compositing;
— Circulation indices;
— Specification.

All synoptic classifications are covered by this list, aside from a couple of hybrid types (e.g., Schwartz, 1991) and a few relatively uncommon multivariate schemes.

Goals of synoptic climatology

Synoptic climatology has one overriding goal: to understand the relationships between the atmospheric circulation and the surface environment. As discussed above, synoptic climatologists primarily classify atmospheric structures (that is, synoptic types or map patterns) to make climatological sense of the circulation. They relate these structures to the surface environment using statistics. This traditional approach is increasingly being complemented by process-oriented research that links the atmosphere and environment via deterministic models. Still, classification is central to synoptic climatology and is the major focus of this book.

Synoptic classification also has specific goals. The first is to identify the recurring map patterns or variable clusters that typify significant modes of the atmospheric circulation. A second, related goal is to classify each pattern or cluster into one of these modes. Unfortunately, some patterns are atypical and simply do not fit even the most detailed classification scheme, although they may be climatologically significant. For instance, tropical storms do not reach central Pennsylvania often, so these rare synoptic structures do not feature in any synoptic classification of that region. However, the Susquehanna River basin received more than 250 mm of precipitation from 1972's Hurricane Agnes, causing one of the most expensive natural disasters in American history. Evidence suggests that such tropical excursions generally occur once or twice per decade, and, although they seldom pack the punch of an Agnes, they do have a large impact on the region's hydrology.

Automated synoptic classification, one of the most important advances of the last 20 years, has its own additional goals. The first is to generate reproducible synoptic categories. The results of manual classifications are unique because the the investigator bases the categorization on his or her subjective interpretations. Research shows that no two interpreters will produce the same results. Furthermore, the same investigator's "internal" classification will meander over time. A computer will always produce the same results if the investigator sets the algorithm's parameters to the same values, uses identical rounding procedures, and does not change the data.

The second goal of the automated procedures is to reduce the time and labor spent on the classification process. Manual techniques are labor intensive and take considerable time. Computer-based classifications can take comparable amounts of time to develop or adapt to a research site. Nevertheless, when the system is operating smoothly and the investigator becomes familiar with the procedure, automated classification is fast.

Automated classification's third goal is to be objective. Manual classifications are subjective by nature; indeed, they are called "subjective classifications" in many quarters. In contrast, computer-assisted synoptic climatology was referred to as "objective" for many years, and this misnomer persists. Chapters 3 and 4 emphasize the subjective decisions investigators must make which, in turn, determine the results of the automated synoptic classification. In short, computer-based synoptic climatology is inherently subjective.

Key theoretical questions

Synoptic climatology seeks to establish the relationships between the atmospheric circulation and the surface environment of a region. Some of the more important theoretical questions integral to these relationships are these:

— Under what atmospheric-circulation regimes do major changes in the surface-environment response occur?
— Which circulation-forced environmental changes maintain themselves over long periods and which are ephemeral?
— At what speeds do atmosphere-to-surface effects propagate through the climate system?

— Associations exist between the largest scales of atmospheric behavior and regional climate. At the same time, associations exist between the smaller scales of atmospheric behavior and regional climate. What are the relationships between these two sets of associations? What determines these connections?

— How will human activities change the relationships between the circulation and the surface environment in the future?

Answering these questions requires understanding of complex non-linear systems and scale linkages, two of science's more difficult conundrums.

Assumptions

Synoptic climatology, like all branches of science, rests on a foundation of assumptions. The most fundamental of these is that *the atmospheric circulation is a critical determinant of the surface environment*. Before starting a study, the synoptic climatologist must determine that observation or theory suggests circulation–surface links. Otherwise, there is little point in continuing the investigation.

Because many synoptic climatologies rely on surface synoptic charts for their assessment of the atmospheric circulation, investigators assume that *the Bergen school conceptual model of the structure and evolution of midlatitude synoptic-scale cyclones is correct*. The model does capture some basic features of cyclone evolution, but Mass (1991) points out that theory and practice reveal serious deficiencies. These shortcomings are compounded by the lack of uniform and specific procedures for defining fronts and analyzing synoptic charts. He calls for an improved conceptual model, plus clear and consistent guidelines for applying analytical methods.

Each synoptic climatology involves the classification of the circulation, and this assumes that *the atmosphere can be partitioned into discrete, non-overlapping intervals*. Truthfully, the atmosphere is a multi-dimensional continuum, and one synoptic class grades from one category into another. The synoptic climatologist is really looking for important clusters in multi-dimensional space. An important source of variance in synoptic climatology results from the inevitably arbitrary divisions between categories. The synoptic climatologist must set these boundaries so they minimize within-group variance and maximize between-group variance.

Synoptic climatologists must also assume that *the classification identifies all important map patterns or synoptic types*. With this assumption, the personality of the investigator becomes important. "Splitters" do not tolerate uncertainty and ambiguity; they prefer to place every potentially important synoptic structure in a well-defined class. In the example of central Pennsylvania's tropical disturbances, a "splitter" would include a tropical category. "Lumpers" accept more uncertainty and ambiguity; they are happy to work with fewer, more diverse synoptic categories. A "lumper" either would not classify the tropical disturbances or would create a classification which would accommodate this synoptic situation in a broader category. Thus, although this assumption is implicit in all synoptic climatologies, the interpretation of just how many categories are important varies widely among investigators.

From the preceding two assumptions, a corollary assumption comes forth: *the classification methods really do what the investigator thinks they are doing*. Gould (1982) demonstrates convincingly that classification based on statistical software and computer analysis is a black-box approach. Not every synoptic climatologist should have to become an authority on the mathematics and theory of principal components analysis, for example, to use that approach to classification. Each must assume that the so-called "experts" have worked out the bugs and that a given classification procedure is sound. However, this does not relieve the investigator of the responsibility to make informed decisions based on previous research in synoptic climatology. Two related objectives of this book are to sensitize synoptic climatologists to the subjectivity of automated classifications and to make them aware of the critical junctures where investigator decisions are necessary.

An assumption of all synoptic climatologies is that *the temporal scales of the observations and the atmospheric-circulation processes match*. For instance, investigators assume that the diagnostic features on a daily weather map operate on time scales of days. This is a fair assumption. One earth day is approximately 10^5 seconds (86,400 to be precise). One million (10^6) seconds equal 11.6 earth days which is near the extreme outer limit for the longest-lived synoptic-scale systems. Typical synoptic systems exist for several days and, therefore, operate within the same temporal order of magnitude as daily synoptic data. Still, each system varies in the duration of its life cycle and in its forward speed. Moreover, large differences exist between the vigor of summer and winter systems as well as one winter's systems and those of the next. These variations influence the results of any synoptic climatology and are an inconstant source of variance in the data.

Similarly, in automated synoptic climatologies, investigators assume that *the spatial scales of the gridded data and the circulation coincide*. This is not a problem when atmospheric features are larger than the horizontal dimensions of the grid. However, when the grid scale is larger than the size of the atmospheric feature, high-wavenumber features are effectively filtered (for example, Yarnal, 1984a). This can have a tremendous impact on the results of a synoptic climatology. The best example of this filtering comes from polar-low research (for example, Yarnal and Henderson, 1989). Polar lows are small synoptic-scale storms that form over polar waters and can develop hurricane-like characteristics and intensities. They are frequent invaders of polar, and even mid-latitude, coastal zones, posing a considerable hazard to these regions. However, not much was known about these storms until recently because the horizontal scale of these systems is far smaller than that of the upper-air networks over oceanic regions. Now, satellite coverage alerts forecast meteorologists to their presence, and long-term image analysis promotes climatological studies of polar lows. In summary, synoptic climatologists must make sure that the grid scale they are using is commensurate with the atmospheric features of interest.

The last assumption of synoptic climatology is that *within-group variability is not a problem*. Unfortunately, within-group variability *is* a significant problem. Variations within synoptic classes are inevitable for many reasons, including those associated with the assumptions mentioned above; that is, assigning discrete intervals to data residing on a continuum and mismatched temporal and spatial scales.

Even when the investigator meets all assumptions, processes internal to the climate system will increase the variance. For example, most of the contiguous United States suffered a massive drought during the spring and summer of 1988. Yet, a synoptic classification of the eastern half of the country demonstrates that that surface synoptic-type and map-pattern frequencies diverged less from the average than other recent years. In central Pennsylvania, surface ozone concentrations in rural areas were well above normal, but these cannot be explained by increased frequencies of the most polluted synoptic types (Comrie, 1992a). Instead, the ozone concentrations of all types were elevated; in other words, investigators observed within-group variation of ozone. Why? First, surface energy and mass fluxes were very different upstream in 1988. In addition, the upper-level support for surface systems was atypical: a persistent ridge sat over the region. Given this dramatically different thermodynamic regime (sources of moisture were cut off and upper-level motions were suppressed), it was unlikely that the synoptic classes would behave normally. To illustrate, cold fronts normally bring strong convection and precipitation to central Pennsylvania. In the spring and summer of 1988, although their frequencies were near average, these severely weakened fronts yielded no precipitation. The normal cleansing of ozone by precipitation did not take place; after the passage of a cold front, ozone levels stayed high. Thus atmospheric patterns can be internally diverse, and this diversity will cause within-group variation. A difficult but important challenge for synoptic climatology is to develop ways to identify and extract within-group variance. Furthermore, as I did here, investigators must use it to explain the synoptic climatology of an area.

Problems

Apart from the hurdles presented by methodology, synoptic climatologists face a few profound problems that go to the core of their field. In the following, I synopsize some of these difficulties.

First, it is possible to restate synoptic climatology's assumptions negatively.

— The atmospheric circulation is not a critical determinant of the surface environment.
— The Bergen school conceptual model of the structure and evolution of midlatitude synoptic-scale cyclones is not correct.
— The atmosphere cannot be partitioned into discrete, non-overlapping intervals.
— The classification does not identify all important map patterns or synoptic types.

— The classification methods really do not do what the investigator thinks they are doing.
— The temporal scales of the observations and the atmospheric-circulation processes do not match.
— The spatial scales of the gridded data and the circulation do not coincide.
— Within-group variability is a problem.

Although technically correct, these statements certainly exaggerate the case. My point is that in any synoptic climatology, the investigator violates each of the assumptions to some degree. The investigator cannot eliminate any of these transgressions completely; they are native to classification, the observations, and the climate system. Still, synoptic climatologists must be mindful of these shortcomings and wary of violating these assumptions more than is necessary.

There are other problems associated with classification of the atmospheric circulation. Many investigators are unwilling to recognize the inherent subjectivity of automated synoptic climatology (see Johnston, 1968). They tend to treat the output as inviolable, which prevents them from tinkering with the procedure to improve the output. Throughout this book, I take pains to demonstrate that there is nothing sacred about computer-assisted classification. As in a manual classification, investigators should force the output to correspond to their knowledge of a region's climatic environment. A related problem is the unwillingness of many investigators to admit that automated techniques are not "more objective" than manual types. A technique is either objective, or it is not.

As I noted in my introductory remarks, synoptic climatology has much to offer scientists from other fields. However, synoptic climatologists have difficulty communicating the value of their work to non-specialists: investigators from the environmental sciences, the other branches of the atmospheric sciences, and, for those from geography, the other fields of geography. Although it is true that classification provides an intellectual shorthand, synoptic climatologists are guilty of writing and thinking in a jargon-filled shorthand which makes their work opaque to most non-specialists. This includes dynamic climatologists and meteorologists, who have no trouble grasping the concepts of synoptic climatology if conveyed in language they can understand.

A related danger is that the jargon of synoptic climatology obscures the depth of the research to all but the most knowledgeable specialist. Many examples of specious synoptic climatology are present in the literature. These articles were published because reviewers were unfamiliar with the terminology and concepts of synoptic climatology and, instead of returning the manuscript to the journal editor and admitting they were unsuited to the task, they let a bad piece of work get into print. Synoptic climatology is not remarkable in the number of second-rate papers it produces, but I believe that an unusually high proportion of poor research is published because of this impediment to quality control. Unfortunately, this problem hurts synoptic climatology's stature in the scientific community.

Synoptic climatology's deductive, statistical approach to science is alien to many scientists, especially meteorologists. Some are truly bothered by this approach; they believe that the synoptic climatologist's tendency to mask the complex feedbacks and non-linear relationships of the climate system in a black box is poor science. Others simply do not understand it. Today's meteorological training is so steeped in inductive reasoning that many meteorologists cannot conceive of another way to approach science. A major problem for synoptic climatologists is to prove the scientific validity of their work and to demonstrate the direct benefit of synoptic climatology to meteorology. This has considerable practical importance because most research funding for climatology comes from organizations headed by meteorologists.

The working definition of synoptic climatology states that it seeks to relate the atmospheric circulation to the surface environment. Scale linkage is one of the most intractable problems in science, yet it is by definition central to every synoptic climatology. Analysis of scales is a fast-growing, multidisciplinary field making rapid theoretical and practical progress (for example, Cushman *et al.*, 1988; Hall *et al.*, 1988; Meentemeyer, 1989; O'Neill *et al.*, 1989; Rosen, 1989; Turner *et al.*, 1989). Synoptic climatologists must keep up with and contribute to this accelerating literature. Another related problem that has haunted scientists for decades is the definition of *region*. Well-developed theories of geographical regions and of regional types do not exist, although there has been some discussion on the logic of regionalization (for example, Grigg 1965, 1967). Synoptic climatologists concern themselves more with classification than with the entity to which they are relating their classification. Synoptic climatologists could use the research from regional ecological risk assessment (for example, Suter, 1990; Graham *et al.*, 1991) to develop a better understanding of region and of point-to-region scale linkages in the surface environment.

To summarize, there are several obstacles endemic to synoptic climatology. The assumptions of the field are impossible to obey. Synoptic climatologists are somewhat removed from today's scientific mainstream and must do a better job communicating with their fellow scientists. Even the definition of the field targets a goal that is one of science's most difficult questions.

Book rationale and structure

Rationale

Perhaps the biggest problem in synoptic climatology is that it has no clearly articulated theory, methodology, or sense of purpose. Barry and Perry (1973) attempted to develop this foundation for the field, but their timing was bad: the advent of computer-based classification and analysis quickly dated their work. In the two decades since then, synoptic climatology has remained a loose collection of related research with little coherent direction.

Synoptic Climatology in Environmental Analysis has several goals. The first is to propose a simple, straightforward set of definitions, methods, goals, assumptions and problems inherent to synoptic climatology. I presented this theoretical framework in the preceding section.

The second goal is to review the literature. Hundreds of papers, reports, and monographs on synoptic climatology have been published since Barry and Perry. Except for periodic reviews in *Progress in Physical Geography* (Barry, 1980; Perry, 1983; Smithson, 1986, 1987, 1988), no single reference on the recent literature is available. Of the published sources, none clearly organize the literature on a theoretical or methodological basis.

The third goal of this book is to furnish a guide to synoptic-classification methods. With the explosive growth in computer-assisted classification, students of the field have grown increasingly confused by the diverse methodological strategies. No authority plainly separates these classification techniques. For instance, what are the similarities or differences among the dozens of papers that use eigenvector analysis for synoptic classification? It is difficult to differentiate the many different classification strategies without a careful, systematic review of the literature. Thus, it is natural to review the literature (goal 2) according to the type of method used (goal 3).

Another shortcoming in the synoptic-climatological literature is the lack of instruction on how to use the various classification methods. There is no primer on the subject with the result that teaching or learning synoptic classification is difficult. This dilemma is compounded by the fact that the number and types of classification methods are uncertain. With this in mind, the fourth goal of this book is to provide practical instruction on how to conduct a synoptic climatology through step-by-step descriptions, worked examples, and research scenarios.

Investigators have no basis on which to judge which synoptic-classification method is most suitable to their application. A few head-to-head comparisons have appeared in the literature over the years (for example, Blasing, 1975; Ladd and Driscol, 1980; Key and Crane, 1986), but no study has comprehensively compared the various classification techniques. This is understandable because undertaking just one synoptic climatology is a big job; conducting several at once is a monumental task. Nevertheless, the fifth goal of this book is to evaluate and compare the performance of the sundry classifications. Using a suite of model-performance statistics, I quantitatively evaluate the overall performance of a number of popular synoptic classifications and compare their relative performances. I also rate model performance in relation to four scenarios (urban air quality, acid rain, agriculture, and fluvial hydrology) to show that results vary according to the environmental setting.

Finally, synoptic climatology is an applied science. Many surface variables are at least in part controlled by variations in the atmospheric circulation. Scientists from all environmental disciplines can benefit by identifying these controls and evaluating their data by synoptic class. Unfortunately, only a few are aware of synoptic climatology and its advantages. Most of these are put off by a wall of jargon and, worse, a disorganized field with no recognizable theory and methodology. Thus, the sixth goal of this book is to show environmental scientists of all stripes that synoptic climatology can be useful. I hope to make synoptic climatology accessible to all by providing a simple theoretical framework, a sensible methodological structure, worked examples of the main classification techniques, appraisals of which classifications work best, and applications to important environmental problems.

Structure

This book is primarily structured around the synoptic-classification techniques. Chapter 2 covers manual classification. First I introduce the classification methodology. In the case of the manual techniques, this requires little discussion; for most of the methods covered in subsequent chapters, this discourse takes several pages. Next I review previous studies that used manual classification. Because this literature is large and diverse, I break the review into sections based on the specific form of the manual classification approach. I follow the review with the worked example of manual classification. Here, I show the steps the investigative team took to produce an original manual classification of daily weather maps. I conclude the chapter by presenting the results of the worked example, including mean weather maps, annual frequencies, seasonal variations, and interannual fluctuations.

The automated methodologies of Chapters 3, 4 and 5 follow the basic plan initiated in Chapter 2. Methodology, literature, and worked examples are presented in order. Chapter 3 addresses correlation-based map-pattern classification. Chapter 4 presents the eigenvector-based techniques, including synoptic typing, map-pattern classification, and regionalization. I also mention other multivariate statistical schemes in a short section of this chapter. I only give worked examples of eigenvector-based synoptic-type and eigenvector-based map-pattern classification. Chapter 5 acquaints the reader with three diverse classification techniques: compositing, indexing, and specification. Only compositing and indexing show worked examples.

Chapter 6 focuses on the environmental scenarios. Four scenarios, with a total of five variables, are employed. The urban air-quality scenario uses surface ozone data from Pittsburgh, Pennsylvania, while the acid-rain scenario studies sulfate concentrations in precipitation collected at the Pennsylvania State University (Penn State). Corn (maize) yields in southwestern Pennsylvania represent the agricultural scenario. The fluvial-hydrology scenario features two variables: runoff from a small watershed in central Pennsylvania and associated in-stream sulfate concentrations. In each scenario, the synoptic classifications presented in the worked examples of Chapters 2, 3, 4 and 5 are related to the environmental variable or variables.

Before addressing the scenarios, I introduce a suite of model-comparison statistics. These quantitative tools permit me to evaluate the overall skill of the synoptic-class frequencies in predicting the value of the surface environmental variable. Furthermore, they provide the means for assessing the relative ability of each classification technique to relate to the surface environment. In other words, they show which synoptic classification does the best job in each environmental scenario.

For each scenario, I describe the environmental setting, problem and variable(s); then I reveal the performance of each of the classifications; and finally I present a scaled-down synoptic climatology using the classification that scored best. In one scenario, none of the classifications fared well, but I show an alternative synoptic approach that does permit the investigators to complete the synoptic climatology successfully. I end Chapter 6 by summarizing and discussing the results.

In Chapter 7, I address the future of synoptic climatology. I pin tomorrow's hopes on interdisciplinary collaboration and theoretical and methodological advances in the field. For now, however, I concentrate on the methods of today.

2 Manual classification

The first method for classifying the atmospheric circulation predates computers and relies upon the investigator's knowledge and judgment. In these classifications, the investigator subjectively groups all synoptic data into predetermined categories. This strategy is often called "subjective." Research on the nature of computer-assisted synoptic classifications, however, demonstrates that automated procedures are also inherently subjective. For this reason, I refer to non-automated classifications as *manual*.

The two approaches to manual classification are the two approaches to synoptic climatology: environment-to-circulation and circulation-to-environment (Figure 1.1). Each is straightforward. In either approach, the investigator starts the classification process by assembling the circulation data, often a collection of weather maps. Surface environmental data for the second stage of the synoptic climatology should be collected at this time, too.

If the investigator first manually classifies the atmospheric circulation and then relates it to the environment, this is a circulation-to-environment classification. For instance, the investigator might want to study heavy precipitation events over a region using weather maps. Previous research suggests that high precipitation totals may be associated with southerly flows and cyclonic curvature of the pressure patterns. Thus, setting these as the two principal criteria, the investigator manually classifies the weather maps, sorting them into periods that have southerly flow and cyclonic curvature over the region and periods that do not. Although the investigator is using knowledge of the probable circulation–environment association, the environmental data are not controlling the classification of the circulation data. The investigator has no idea whether any given day with cyclonic curvature and southerly flow is precipitating or not. Thus, the classification of circulation data is independent of the environmental response.

If, however, the investigator wants to use the surface environment to control the manual selection of the circulation data, then an environment-to-circulation classification is in order. Continuing with the precipitation problem of the previous paragraph, the investigator decides that he or she only wants to classify those days on which heavy precipitation fell. The circulation data are separated manually into two piles: one matching heavy-precipitation days and one that does not. Further study of the circulation data may confirm that cyclonic curvature and southerly flow relate directly to intense precipitation events. The point here is that the classification was not independent of the environmental response; indeed, the classification process was constrained by knowledge of the surface environment.

Although automated classification procedures are popular, many investigators still favor manual classification techniques for at least three reasons. First, the investigator can produce a manual classification without a computer. This is a benefit in developing or former-socialist countries where computers may not be available or of sufficient power for classification. Classifying without a computer is also an advantage to older investigators who are not comfortable with computers but who "grew up" with manual classifications and bring considerable knowledge, experience and intuition to the classification process. Furthermore, the training of new synoptic climatologists should begin with manual techniques. Trainees are usually anxious to move immediately to the quicker, easier, flashier and, therefore, more seductive automated techniques. Unfortunately, if they are allowed to start their synoptic-climatological education on the computer, they often come away frustrated by their inability to understand the results, both in terms of the classification process and the physical basis of climate. Learning to classify circulation data manually gives young investigators an appreciation of the difficulties inherent in the classification process, an awareness of the regional modes of atmospheric variation, and a sensitivity to the linkages between atmospheric processes and the surface environment. Once the trainees have this feel for synoptic climatology, the time is right to try computer-based classifications.

Second, the investigator can tailor the manual classification to the exact needs of the data and research. In subsequent chapters, I demonstrate that there are limits to the kinds of data computer-based classifications can handle and the results they can produce. In short, computer-based classifications dictate the form of the data and results. In contrast, manual classifications can use diverse and unusual kinds of data, including those data not amenable to a digital format. The investigator can shape the manual classification so that the final results meet the needs of the research. Manual classification is infinitely flexible.

Finally, and related to the second reason for manual classification's popularity, the investigator can control the classification process completely. No matter how familiar the investigator is with its mathematical underpinnings, or how much the parameters are fine tuned, there will always be a "black box" element to automated classifications. In contrast, at every step of the manual classification process, the investigator consciously or subconsciously manipulates the classification to match his understanding of the physical world and to produce the type of results required by the research design.

Not all manual classifications, however, are designed for a specific research project. Important generic manual classifications have been developed for general use. In this chapter, I base the literature review on the distinction between generic manual classifications and unique classifications.

Despite the positive features of manual classification, this strategy suffers two major drawbacks. First, manual classification is labor intensive by definition. One investigator, who wishes to maintain complete control over the classification process, can shoulder the work. Alternatively, a team of investigators drilled on a common training set can share the effort. No matter how the load is divided, however, the process requires many person-hours. This might not be a problem under some circumstances. For instance, if the investigator intends to use the classification on several occasions, the investment of time comes with the first application. After that, there is no further time required for classification unless updates are needed. To keep this in perspective, setting up an automated procedure also takes considerable time before the first application; its payoff similarly comes with subsequent applications. If the manual classification is an educational tool, either training new synoptic climatologists or teaching an experienced investigator about a new region, atmospheric scale, or set of variables, then the large expenditure in time is also worthwhile.

The second major drawback of manual classifications is that they cannot be duplicated. Even with well-conceived, well-defined classes, a large proportion of all synoptic data is borderline or ambiguous and the investigator must make a subjective decision on the categorization of the data. Each investigator will vary in his or her interpretation of these equivocal data. Thus, no matter how well trained investigators are and how rigorously they apply the classification rules, the results will not be consistent among investigators. Worse, each individual will vary in his or her interpretations over time. Therefore, the results of a manual classification cannot be reproduced. This contrasts starkly with computer-assisted classification, in which results can be duplicated.

These drawbacks are so significant that many investigators have turned their backs on the immense potential of manual classification. This is a mistake. A well-balanced, integrated synoptic-climatology program should use manual methods alongside automated procedures.

In this chapter, I review the manual classification literature, discussing the various classification techniques and their applications to surface environmental problems. Then I introduce a worked example of a manual classification procedure, from conception through implementation and including problems encountered in the classification process. I conclude with the results of this classification, which I apply to the environmental scenarios in Chapter 6.

Methods and applications

The manual classification literature is vast and diverse, overwhelming the fledgling synoptic climatologist and more experienced reviewer, alike. To make it more manageable, I split the literature into two groups: generic classifications and unique classifications. Besides utilitarian concerns, this tactic has conceptual advantages. Studying just a few important generic schemes and their varied environmental applications allows the reader: (1) to digest the structure of the classification; and (2) to grasp the linkages between atmospheric processes and the surface environment. In contrast, the unique classifications are too numerous and dissimilar to allow the reader a chance to ponder them. The unique classifications, however, are important to review because they introduce the reader to a rich assortment of classification stratagems and environmental applications.

From both of these discussions, I hope to develop in the reader an appreciation of synoptic climatology in general and manual classification in particular. The knowledge from this chapter forms the foundation upon which I build the more technical discourses of subsequent chapters.

Generic classifications

Generic circulation-to-environment classifications are designed so they can be related to many surface environmental situations. This is both a strength and a weakness. It is a strength because the investigator does not need to spend time on classification; instead, he or she can focus on the more important issue of relating the surface environment to the atmospheric circulation. Also, over time, investigators working in one region associate a single generic classification with many environmental variables. This produces a holistic appreciation of that region's synoptic climatology and, furthermore, makes it possible for investigators to explore simultaneous associations among environmental variables and the atmospheric circulation. The main weakness of the generic schemes is common to all circulation-to-environment classifications: investigators cannot optimize the relationship between the circulation and the surface. Users of generic manual classifications overlook this weakness for the convenience of having a reliable, ready-made classification at their disposal. If it is critical to explain the most variance possible in the relationship between the surface and the synoptic categories, then the investigator should consider designing an environment-to-circulation scheme such as a unique manual classification.

A decisive test of the worth of a generic classification is its continued application to environmental and climatological problems. In the last 20 years, some early schemes, such as the Dzerdzeevskii or Wangenheim-Girs classifications, have only been used occasionally (for example, Knox, *et al.*, 1975; Kozuchowski and Marciniak, 1988). Automated procedures based on commercially-available atmospheric data sets, such as those I discuss in Chapters 3, 4 and 5, have replaced these classifications. Other generic classifications have received limited use since their inception (for example, Barry *et al.*, 1975; Altshuller, 1978). Still, investigators continue to develop and apply generic classifications (for example, Flocas, 1984; Flocas and Giles, 1991).

There are several highly regarded manually derived synoptic classifications available in the literature. These generic classifications have been applied to a wide range of environmental problems. In this section, I discuss the Lamb weather types, Muller classification, Grosswetterlagen, and cyclone model.

The Lamb weather types. The noted British climatologist H.H. Lamb developed a synoptic-scale, daily weather-map classification for use over the British Isles. The original catalog extends from 1861 to 1971 and is updated continually in the University of East Anglia's *Climate Monitor*. Because

Lamb has been the only analyst to classify the data, this long synoptic record is internally consistent.

Lamb (1972) recognized seven basic types.

A. *Anticyclonic* days feature highs centered over, near, or extending over the British Isles. They are dry with light winds and are warm in summer and cold or very cold in winter.

C. *Cyclonic* types have depressions centered over Britain or Ireland at sometime during the day. These types are associated with wet or disturbed weather, with highly variable wind directions and strengths. Cyclonic days are relatively mild in fall and winter but relatively cool in spring and summer.

W. *Westerly* conditions occur when high pressure is south of the British Isles and low pressure is to the north. The resulting zonal flow steers sequences of depressions and ridges across the Atlantic and over Britain. Weather is unsettled and changeable, with winds veering rapidly between south and northwest, and rain falling on the western and northern districts. Westerly days are cool in summer, and mild in winter with frequent gales.

NW. *Northwesterly* types find the subtropical anticyclone displaced either northeast toward the British Isles, or northward, off the west coast of the islands. Depressions are steered around the clockwise flow, traveling into the region from the northwest. Weather is unsettled or changeable, particularly in the north. Days associated with NW are cooler than the W-type and milder than the N-type.

N. *Northerly* days have high pressure to the west-northwest of Britain, with low pressure over the region of the North Sea, the Baltic Sea and Scandinavia. Depressions are directed southward, producing cold, disturbed weather in all seasons. Northerly types are snowy in winter and are associated with late spring and early autumn snows and frosts. Strong northerly winds often usher in N-days.

E. *Easterly* types spotlight anticyclones stretching from Scandinavia toward Iceland. Depressions are restricted to the western North Atlantic and the vicinity of the Iberian peninsula. Winds flow off the Eurasian land mass, bringing cold weather in autumn through spring, with occasional snow in eastern Britain. This type is warm and sometimes thundery in summer.

S. *Southerly* flows result from high pressure over central and northern Europe. Depressions off the Atlantic are blocked west of the British Isles. S types are warm and thundery in spring and summer and mild in autumn. They are cold in winter when the air flows off the continent, but mild when the trajectory is over the ocean.

Lamb also identified unclassifiable days and hybrid types. Unclassifiable days occur when patterns are (1) chaotic with weak flow, (2) change quickly during the day, or (3) form "incompatible" hybrids. Hybrid types satisfy the definitions of more than one individual type. They reduce the number of unclassifiable days. Also, Lamb thought that inclusion of these types makes the technique more objective, preventing the forcing of maps into ill-fitting classes. With the basic, hybrid, and unclassifiable categories, the scheme has a total of 27 types (Table 2.1).

Table 2.1 The Lamb synoptic types

Basic types	Hybrid types
A	
	ANE
	AE
	ASE
	AS
	ASW
	AW
	ANW
	AN
	NE
E	
	SE
S	
	SW
W	
NW	
N	
C	
	CNE
	CE
	CSE
	CS
	CSW
	CW
	CNW
	CN
Unclassifiable	

Lamb (1972) suggested that when simplicity is appropriate for statistical analysis, the hybrids can be converted back to the basic types. A CW-type, for example, becomes C- and W-types, with each CW-day contributing one-half day to the C and W categories. Jones and Kelly (1982) demonstrated this idea more elegantly with principal component analysis. They found that six of the 27 categories accounted for most of the temporal variance in the Lamb weather-type catalog, and these were basic, rather than hybrid, types.

Additionally, they showed that the three most common weather types, W, A and C (Table 2.1), accounted for 57% of the variance. Briffa *et al*. (1990) extended this idea further, showing that the principal components can be used to fashion four generic indices of climatic variation over the British Isles. These indices highlight the clear decline in the westerlies experienced in the last several decades. This decline is marked in all seasons, compensated for by increases in more cyclonic and anticyclonic types, and has become even more pronounced in the 1980s.

Several investigations have related the Lamb synoptic types to temperature and precipitation. Sowdon and Parker (1981) studied the association between variations in central England air temperature and the Lamb types. They determined that, in a general sense, circulation changes account for observed temperature changes. Yet, the circulation changes themselves are not always reflected in the synoptic-scale Lamb-type frequencies, thus limiting the strength of the temperature–type relationship. Jones and Kelly (1982) also found weak but physically reasonable correlations between their principal components and central England temperatures. In contrast, they discovered that the first two components accounted for 63% of the variance in English and Welsh rainfall. In a more detailed investigation of relationships between the Lamb types and English and Welsh rainfall, Wigley and Jones (1987) concluded that high (low) precipitation strongly associates with cyclonic (anticyclonic) types. Additionally, westerly types explain much of the precipitation variance along the west coast. Taken together, the results of the above studies showed that the dominant control on precipitation in England and Wales is the atmospheric circulation, although orographic effects modulate this in west coastal regions. Temperature is weakly related to the Lamb types. Evidently, factors other than the synoptic-scale circulation exert more control on this variable.

In an interesting application to wind-energy potential in Britain, Palutikof *et al*. (1987) looked at relationships between the Lamb types and wind speed. They compared output from principal components analyses of wind speed and the Lamb types, using monthly wind data from 52 British stations and the results of Jones and Kelly (1982). They found that the first principal component of each analysis explained much of the data's variance: 51% for wind and 34% for the Lamb types. Regression of the component scores from the Jones and Kelly analysis with the annual mean scores from the wind analysis produced a correlation coefficient of -0.902. Therefore, given the strong positive loading of the A category and strong negative loading of the W category on the synoptic-type component, these factors suggest that decreased frequencies of anticyclonic types (and concomitant increases of westerly types) associate with stronger winds over Britain. Low wind speeds relate to increased frequencies of anticyclones and decreased westerlies. There is, however, a spatial element to this relationship that is strongest at west-coast locations and weakest in the east. Palutikof *et al*. (1987) neglect to inform the reader if there has been any time in the past 122 years when anticyclonic dominance has been so great that wind-power generation along the west coast would have failed.

The Lamb weather types have also been applied to the problem of acid rain. Davies *et al*. (1986) related the Lamb catalog to precipitation acidity in Eskdalemuir, southwestern Scotland. They observed that the two types that produce the most precipitation, C and W, have different acidities. The most acidic rainfall episodes at Eskdalemuir are associated with C-conditions, with this type accounting for 20–25% of the station's hydrogen-ion deposition. W-types are about half as acidic. Airflow associated with cyclonic types often passes over industrialized areas of England or the continent, while westerly types strike Britain after passing over the open Atlantic. They concluded that because interannual variations in synoptic-type frequencies control much of the variance in acid rain at Eskdalemuir, climatic variation must be taken into account in assessments of acid–rain trends. Davies *et al*. (1990) demonstrated that the annual acidity of rainfall at Eskdalemuir is in large part controlled by the frequencies of Lamb synoptic types. Most important is the dilution caused by westerly types: the more westerly types in a given year, the lower the annual precipitation acidity. Thus, for the period 1978–84, they attributed trends in the precipitation composition to the annual numbers of C- and W-types. In a follow-up study, Davies *et al*. (1991) showed that the relationship between anthropogenic ion content of precipitation and the Lamb weather types varies with geography and relationship to emission source. In relation to the information for Eskdalemuir given above, they found, for example, that C-types produced relatively low acidities in Goonhilly, Cornwall. Compared to Eskdalemuir, C-types approach Goonhilly from the open ocean and are free of anthropogenic pollution; in contrast, Stoke Ferry, East Anglia is so near the major source areas of pollution that it can be considered within them. No one synoptic type stands out as an important bearer of pollution at Stoke Ferry; all rainfall has relatively high anthropogenic ion counts. Thus, the regions that are most sensitive to anthropogenic ion variations caused by fluctuating synoptic-type frequencies are those more peripheral, not proximal to the sources of pollution.

Heathcote and Lloyd (1986) use Lamb weather types to explain variations in the isotopic composition of daily rainfall in Lincolnshire. They found the lightest isotopes in those weather types with air masses originating over the relatively cool waters of the northern North Atlantic. The heaviest isotopes are associated with Lamb types either of continental origin, perhaps because of previous rainout over Europe, or originating over the subtropical North Atlantic, where the higher evaporation temperatures enrich the isotopic composition of the water vapor.

The Muller classification. R.A. Muller of Louisiana State University devised a generic manual synoptic-type classification explicitly to relate to the surface environment and human activities of the United States Gulf Coast (Muller, 1977). The classification has eight typical surface weather patterns found over the contiguous United States. Muller gave the eight types descriptive names for easy reference. A brief summary of the synoptic types follows.

Pacific High days have a deep surface low to the north of the Gulf states. High pressure from the western Pacific Ocean or Pacific Coast states brings mild, dry air into the region.

Continental High is the most common type. It has an anticyclone just east of the Rocky Mountains, with northerly flow of continental polar or Arctic air carrying relatively dry and clear conditions to the Gulf Coast.

Frontal Overrunning finds the polar front along the Gulf Coast, with either mild air from the Pacific or cool-to-cold polar (or Arctic) air north of the front. Warm, moist tropical air overruns the front, delivering heavy clouds and precipitation to the Gulf Coast.

Coastal Return types occur when a high-pressure cell extends over the eastern one-third of the United States and an incipient low is forming in the lee of the southern Rocky Mountains. Airflow into the Gulf Coast region has easterly components, with muggy tropical air in summer and in winter mild continental-polar air that has made a short passage over the Gulf.

Gulf Return is the second most common pattern. It has the continental high farther to the east than in the Continental High and Coastal Return configurations and a well-developed low over the eastern Great Plains, northwest of the Gulf Coast. This stimulates the southerly flow of warm, moist air from the Caribbean and Gulf of Mexico.

Frontal Gulf Return is similar to the Gulf Return type, except that the low passes directly over or just to the north of, the region. This conveys warm-sector return flow off the Gulf and weather associated with cold frontal passage.

Gulf Tropical Disturbances affect the Gulf Coast in summer and fall. These events range from weak easterly waves to deadly hurricanes. Rainfall amounts are often huge, with one-day totals sometimes exceeding 250 mm.

Gulf High configurations find the Bermuda high farther south and west than normal, bringing southwesterly flow into the Gulf Coast. Although this synoptic type usually transports maritime tropical air, it can occasionally direct warm, dry continental air into the region from Mexico and the desert southwest.

Muller and his students update the classification monthly and publish a synoptic calendar in the *Louisiana Monthly Climate Review*, available from the Louisiana Office of State Climatology. The classification runs from 1951 to the present.

Besides distinct weather-type properties, the synoptic types show seasonal cycles, interannual variability, and geographic variation of these properties within the region (see Muller, 1977; Muller and Wax, 1977; and Muller and Willis, 1983). Seasonal cycles are illustrated, for instance, by the Frontal Overrunning and Frontal Gulf Return types. They are uncommon in summer and early autumn, when the region is on the poleward margins of tropical easterly flow. During the remainder of the year, the westerlies overlie the area and these types are frequent. Continental Highs and Frontal Overrunning, for example, show considerable interannual variability. Years with many of these synoptic types have more continental–polar air masses and, therefore, lower temperatures and precipitation totals along the Gulf Coast. In those years when these continental types decrease, synoptic types associated with tropical air (for example, Gulf Return, Frontal Gulf Return, and Gulf Tropical Disturbance) increase, with attendant rises in temperature and humidity. Conditions at the New Orleans and Lake Charles, Louisiana airports epitomize geographical variation in the surface response to the various synoptic types. Both are situated next to lakes: New Orleans' airport is southwest of Lake Pontchartrain, while Lake Charles' airport is on the northeast of Lake Calcasieu. When Continental High or Frontal Overrunning bring northwesterly winds across the 25–mile diameter of Lake Pontchartrain to New Orleans, they traverse land-surface areas to reach Lake Charles. Accordingly, these continental synoptic types bring different weather properties to the two locations. In summary, based on their ability to capture the characteristic temporal and spatial variations of the Gulf Coast climate, the Muller types appear to be robust.

Muller, his students, and colleagues have applied the manual synoptic-type classification to a wide selection of environmental problems, including flooding, air quality, insect migration, evaporation, and hourly precipitation. An early application concerned water levels and salinity in the Barataria basin of the Mississippi bayou country. Wax et al. (1978) proved that water levels respond strongly to the meteorological inputs indexed by the Muller types. On the one hand, invasions by northerly flows of dry air caused water levels to fall dramatically; on the other hand, synoptic types that introduced tropical marine air into the basin, with or without precipitation, raised water levels. Relationships to salinity were not as clear. That result was probably degraded by at least two major problems. First, the investigators relied on only one salinity-observing station too close to the Gulf of Mexico; that is, the salinity regime of the open Gulf partially masked the fresh-water response. Second, salinity changes are undoubtedly more complex than water-level variations. Childers et al. (1990) used the findings and classification strategy of Wax et al. to relate climatological forcing of water levels in Louisiana estuaries to variations in the area's shrimp catch.

Muller and Faiers (1984) reviewed the synoptic situations associated with the devastating flood events observed in Louisiana in winter 1982–3. The floods were associated with three Muller types: Frontal Overrunning, Frontal Gulf Return, and Gulf Tropical Disturbance; the frequencies of these types were normal. The precipitation totals associated with these types were high, but nowhere near record setting, as suggested by the floods. Flooding seems to have been more directly related to the saturated soils that were present at the start of each event.

Muller and his students have addressed air-quality issues with the classification. Three of the Muller types are conducive to the formation of high surface–ozone levels in the Baton Rouge, Louisiana area (Goldberg,

1984). Not surprisingly, all depict high-pressure systems (see Comrie, 1990). Successive days of the Continental High synoptic type produce the highest ozone levels, as northwesterly airflow brings ozone precursors from the industrialized area to the northwest, and the increased solar radiation generates ozone. Coastal Return weather is associated with high precursor totals produced by vehicle emissions in eastern and southeastern Baton Rouge. Using the Muller types, mixing heights, and dispersion, Muller and Jackson (1985) evaluated air-quality potential at Shreveport, Louisiana. Four synoptic types — Gulf High, Pacific High, Continental High, and Coastal Return — associate with low-level night-time inversions, high sunshine totals promoting photochemical production of pollutants, and poor air quality. Each transports air through an arc ranging clockwise from west to east. These patterns are present about one-third of the time. Two other synoptic types with winds veering from southeast to southwest, Gulf Return and Frontal Gulf Return, relate to better air quality. They are characterized by higher overnight mixing heights and strong horizontal dispersion, plus cloud and decreased solar radiation during daylight hours. These types also occur about one third of the time. They conclude that, from the perspective of the town's synoptic climatology, the most favorable location for industries producing noxious materials would be southeast to southwest of the city.

Muller and Tucker (1986) applied the Muller types to the problem of moth transport and migration. The investigators found that the many Frontal Gulf Return and Gulf Return days that occur in late winter–early spring bring strong southerly winds. These winds have the potential to transport moths from Mexico into the midwest and Mississippi River Valley. Similarly, the high proportion of Continental High and Frontal Overrunning days in the autumn produce northerly winds and opportunities for moth transport back to Mexico. Their work shows that the potential for applying synoptic climatology to animal transport and migration is tremendous.

McCabe and Muller (1987) compared pan-evaporation estimates based on the frequency of Muller types with calculations produced by three more traditional methods for estimating evaporation. Predictions from synoptic-type frequencies beat the temperature-based Thornthwaite method and the solar radiation-based Jensen–Haise technique, while tying a modified Penman approach. Thus, synoptic-climatological methods could be used for those areas where pan-evaporation measurements are not available but synoptic weather maps are.

January hourly precipitation at Lake Charles varies with the Muller types (Faiers, 1988). Frontal Gulf Return generates the most intense January precipitation, while over three-quarters of the hourly precipitation events are related to Frontal Overrunning. In addition, the frequencies of these types vary from year to year. Years with meridional flow over North America and troughing over Louisiana produced higher numbers of the strong frontal types and, thus, more hourly precipitation events. These, of course, are cooler Januaries. Warmer Januaries are associated with zonal flow and manifest more non-frontal types and lower rainfall totals. This was especially true for the zonal periods up to the mid–1950s and in the early

1970s. In addition, the increased cyclogenesis over the Gulf during El Niño events enhances precipitation at Lake Charles.

Grosswetterlagen. The central Europeans have approached synoptic-climatological classification in a slightly different way. Other generic schemes, such as those of Lamb and Muller, concentrate on classifying daily synoptic charts. Investigators using these generic schemes do, however, consult the maps of several days which precede and follow any given chart in order to place it in a larger spatial and temporal context, thus improving the classification performance. In contrast, the central Europeans classify synoptic regimes with durations of several days first; later they break down these periods into individual days.

The best known of these classifications is the *Grosswetterlagen* of the Deutscher Wetterdienst (German Weather Service). Over several decades, Franz Baur developed the *Grosswetterlagen* concept using surface pressure patterns. Hess and Brezowsky subsequently modified the system to include the upper-air data that became available after World War II. Since 1949, the *Grosswetterlagen* have been determined by using simultaneous surface and upper-air charts; older *Grosswetterlagen* were prepared from surface weather maps only. The most recent catalog (Hess and Brezowsky, 1977) contained daily *Grosswetterlage* from 1881 to 1976. Annual updates to the present are available from the Deutscher Wetterdienst (1977–1990).

A *Grosswetterlage* is a large-scale pressure-pattern complex over Europe and the adjacent northeastern North Atlantic. The basis of the classification is *steering*; the idea that general atmospheric flow, rather than surface wind directions, is critical to the character of large-scale weather. Although *Grosswetterlagen* focus on central Europe (Barry and Perry, 1973), the classification may relate to all surface areas surrounding the Baltic and North Seas. During a *Grosswetterlage*, the main features of weather are almost constant over Europe. Each *Grosswetterlage* typically lasts three days or more, after which there is rapid transition to another circulation type. The durations of the various *Grosswetterlagen* are log-normally distributed, with most categories' medians being about 3.5 days (van Dijk and Jonker, 1985).

Each day is assigned to one of 29 *Grosswetterlagen* or an unclassifiable category (Table 2.2). Many *Grosswetterlagen* have zonal and anticyclonic forms, while others are unique. For convenience, the *Grosswetterlagen* can be broken down into three general categories: zonal, meridional, and mixed. There are four zonal types (about 28% of the days since 1900), 18 meridional types (approximately 36% of the days), and seven mixed types (nearly 35%; Fitzharris and Bakkehoi, 1986). Fewer than 1% of the days are caught in the unclassifiable, "transition" category.

Surprisingly little work that relates the *Grosswetterlagen* to the surface environment has been written in English-language journals in the last 18 years. (I have seen few non-English synoptic-climatology references, as well.) Most have dealt with statistical characteristics of the *Grosswetterlagen* (for example, van Dijk and Jonker, 1985) or temporal variations and their

Table 2.2 The *Grosswetterlagen* types

Zonal	Mixed	Meridional	Unclassified
WA	SWA	NA	U
WZ	SWZ		
WS	NWA	NZ	
WW	NWZ	HNA	
	HM	HNZ	
	BM	HB	
	TM	TRM	
		NEA	
		NEZ	
		HFA	
		HFZ	
		HNFA	
		HNFZ	
		SEA	
		SEZ	
		SA	
		SZ	
		TB	
		TRW	

causes (for example, Klaus and Stein, 1978; Bardossy and Caspary, 1990). Fitzharris and Bakkehoi (1986) tried to explain avalanche winters in Norway in terms of *Grosswetterlagen* frequencies. They found no strong signal and concluded that no one pattern or set of patterns produces dangerous avalanche situations. They suggested that (1) *Grosswetterlagen* may not be suitable to avalanche analysis, (2) Norway may not be sensitive to changes in *Grosswetterlagen*, and (3) sequences of weather types, rather than frequencies, may be more critical to avalanches. Fraedrich (1990) related the variations in *Grosswetterlagen* to the El Niño and La Niña phases of the El Niño/Southern Oscillation (ENSO) phenomenon. He discovered that European weather types, cloudiness, and temperature vary with the phase of ENSO. To understand the regional effects of climate change, Bardossy and Caspary (1990) studied the temporal and spatial associations of *Grosswetterlagen* with temperature and precipitation. Since sometime in the 1970s, they found an increase in the frequency of wintertime zonal types and a corresponding decrease in meridional circulation types. The result of this change is milder, wetter winters in central Europe, with precipitation falling mostly as rain.

The cyclone model.   The classic cyclone model of the Bergen school — with its surface low–pressure center, warm, cold and occluded fronts, and associated anticyclones — forms the basis of the Lamb weather types, Muller classification, and *Grosswetterlagen*. Disregarding the details of the classification schemes, the chief difference among these procedures is their regional focus. Lamb's scheme concentrated on the British Isles, Muller spotlighted the Gulf Coast, and Baur fixed on central Europe. A logical extension of this reliance on the cyclone model is the development of a truly generic manual classification for use in almost any extratropical location.

Several investigators have devised such a model. For instance, Lindsey (1980) developed one for analyzing coastal wind regimes along the United States Atlantic Coast. R.A. Pielke, Lindsey's graduate adviser, modified and embellished that model, applying it to problems of mesoscale forecasting (Pielke, 1982; Forbes and Pielke, 1985) and mesoscale air quality (Yu and Pielke, 1986). At the same time, Barchet formulated a similar classification scheme for studying wind energy in the midwestern United States (Barchet, 1982; Barchet and Davis, 1983). Barchet and Davis (1984) extended that to the entire contiguous United States. I present a third example of the cyclone model later in this chapter for assessment with the environmental scenarios of Chapter 6. Comrie (1992a, 1992b and 1992c) and Comrie and Yarnal (1992) apply this version of the cyclone model to air quality in Pittsburgh, Pennsylvania. Other versions of the cyclone model are presented by EPRI (1983) in their study of atmospheric sulfate variability and by Heidorn and Yap (1986) for assessing synoptic controls on surface ozone concentrations.

The model utilizes a typical wave cyclone at the surface, independent of place (for reference, see the model presented in Figure 2.1). The classifications focus on the typical weather characteristics in the air masses separated by the fronts of the wave cyclone. Each model mentioned in the last paragraph shares the following four categories:

(1) A warm sector ahead of the cold front and south of the warm front, with cyclonic curvature of the isobars, a southerly wind component, and a veering wind (that is, one that is rotating clockwise). Pressure is falling rapidly, and humidity, wind speed, and cloudiness are increasing as the cold front approaches. Temperature is relatively high and precipitation is possible. This configuration is often called *back of high* in the literature.

(2) North of the warm front and spreading westward over the low–pressure center, an area of moderate to high humidity, heavy cloud, and continuous precipitation. The isobars have cyclonic curvature, pressure is slowly falling, and the wind has an easterly component and is backing (that is, it has a counter-clockwise rotation). Thermal conditions vary from relatively cool to moderate.

(3) A sector just behind the cold front with cyclonic curvature of the isobars and a westerly wind component. Cloud cover and humidity are intermediate, pressure is rising, and temperature and humidity are falling. Showers are possible.

(4) Farther west of the cold front and to the east of the high, a region with anticyclonic curvature of the isobars, steady or slowly rising pressure, and northerly wind. Temperature and humidity are relatively low; humidity will continue to fall as the high approaches, but temperature

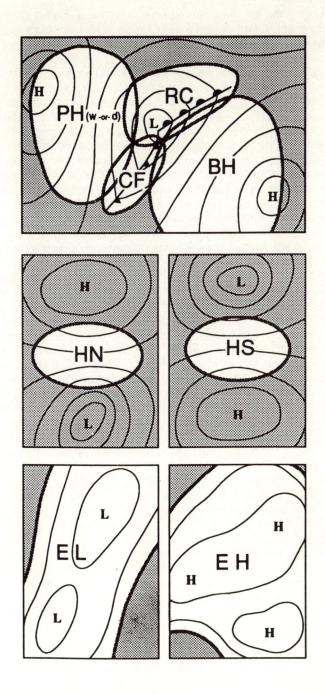

Figure 2.1 The manual classification scheme.

may fall if the air mass is cold or it may rise if the air mass is warm. Compressional and radiational processes under the high may enhance this warming. There is little cloud cover and essentially no precipitation.

Other categories are likely. For instance, some classifications denote a type for the area under the high or just south of the high, preceding or following a cold front. The Barchet scheme identifies a stationary-front category, and a class for occluded fronts and regions west of occluded lows. The strategy I use in this chapter's worked example has categories representing surface patterns under stationary troughs or ridges in the upper troposphere.

There have been several important applications of the cyclone model to environmental problems. For example, EPRI (1983) showed that highest sulfate concentrations consistently occur in back-of-high situations. Sulfur dioxide buildup and conversion of this chemical to sulfate is initiated during the stagnant conditions under the anticyclone. Continued sulfate conversion occurs as the back of high comes overhead. The sulfate is then either (1) transported north, lifted over the warm front, and precipitated out to its north in a prolonged, moderately acidic event or (2) squeezed out suddenly at the cold front in a short, highly acidic precipitation event. (See Yarnal, 1991, for a detailed account of this process in relation to a typical wave cyclone.) Similarly, Heidorn and Yap (1986) and Comrie and Yarnal (1992) found that the back of high is associated with high surface-ozone values in summer. The following make this sector of the wave cyclone ideal for summertime surface-ozone buildup (see Comrie, 1990): high temperatures and relatively clear skies, thus promoting photochemical conversion of the precursors of ozone; sufficient time for photochemical activity to have taken place since the center of the anticyclone passed over head in migrating systems; and frequent stagnation of high-pressure cells, enabling continual transport of ozone and its precursors into one region. In a similar study, Comrie (1992b) demonstrated that specific sequences of cyclone-model sectors produce the highest surface-ozone concentrations in Pittsburgh. Yu and Pielke (1986) established that in the southern Colorado Plateau of the United States, the eastern side of a polar high-pressure system produces stagnant conditions and relatively poor ventilation, especially when the ground is snow covered and the sun angle is low (that is, winter). This creates an inversion lid over the region, increasing the likelihood of diminished air quality from local pollution sources.

In summary, the varied applications using the Lamb weather types, the Muller classification, *Grosswetterlagen*, and the cyclone model demonstrate the value of generic classification schemes in synoptic climatology. Also, they show the potential for synoptic climatology in environmental analysis.

Unique classifications

The literature on unique manual classifications in synoptic climatology is too broad to review on an article-by-article basis. Luckily, like all synoptic climatologies, the unique manual classifications fall into one of two approaches: *circulation-to-environment* and *environment-to-circulation*. In the following brief review of unique manual classifications, I break the unique manual classifications into these two groups. I not, however, address methodology. The classifications are singular by definition and numerous; a project-by-project coverage of classification techniques would be too long and exhausting. Instead, I emphasize applications, providing the reader with a broad overview of the types of problems attacked with unique manual classifications.

Circulation-to-environment classifications. Many investigators have used this unique manual approach to address problems concerning ice and snow. Alt (1978, 1979) created parallel classifications to determine the synoptic controls on the mass balances of the Devon Island and Meighen Island ice caps, located in the Queen Elizabeth Islands of Arctic Canada. She later expanded this work to identify three distinct synoptic situations regulating mass balances throughout the Queen Elizabeth Islands (Alt, 1987). From her work, Alt developed synoptic analogs for application to paleoclimatic reconstruction (for example, Alt, 1983). Similarly, Hay and Fitzharris (1988) associated synoptic weather patterns over the southwest Pacific with ablation on a New Zealand glacier. They discussed their findings in relation to glacier retreat and possible circulation changes during the twentieth century. Stenning *et al.* (1981) showed that formation and maintenance of the katabatic layer over an ablating glacier in southwestern Canada are controlled by the large-scale atmospheric circulation. Circulation-to-environment unique manual classifications were used to establish the relationship between sea-ice extent and cyclone distribution and movement around Antarctica (Howarth, 1983) and over the Bering Sea (Overland and Pease, 1982). Yet, different mechanisms appear to be at work in these two regions. Antarctica's longitudinal variation in cyclone tracks is probably related to the asymmetry of the continent, instead of sea-ice extent. In contrast, interannual variations in cyclone tracks, which appear to regulate the sea ice over the Bering Sea, seem controlled by changes in the global-scale atmospheric circulation.

Several investigators have used circulation-to-environment unique manual schemes to study variations in acid rain. Raynor and Hayes (1981, 1982) demonstrated that precipitation acidity over central Long Island, New York varies with synoptic conditions. Hydrogen-ion concentrations are highest with showers produced by cold fronts and squall lines; total acidity is greatest in the more dilute but more plentiful precipitation associated with warm and occluded fronts. Still, because the relationships between pollutant source region and synoptic patterns vary over time and space, these associations are not fixed. For example, over Northern Quebec, warm fronts produce more acidic precipitation than cold fronts (Singh *et al.*, 1987). This is attributable to the trajectories of air streams associated with these fronts and the time they spent over polluted areas. Over the heavily-polluted region of central Bohemia, Czechoslovakia, Moldan *et al.* (1988) found that cold fronts and fast-moving frontal systems have the lowest acidities, warm fronts have relatively moderate acidity, and that non-frontal and local convective activity produce the most acid precipitation. Again, this suggests that relationships between the source region and air streams are a critical determinant of precipitation acidity.

Circulation-to-environment unique manual classifications have been applied to precipitation variability in many countries of the British Commonwealth, especially the former Commonwealth nation of South Africa. Miron and Lindesay (1983) combined several synoptic classifications to derive a generalized scheme depicting the four main circulation patterns affecting summer rainfall over South Africa. They showed that wet and dry spells have different synoptic-type frequencies and wind regimes. Tyson (1988; see also 1986) formulated a comprehensive classification that explains rainfall over southern Africa on time scales ranging from days to months, seasons, years, and spells of years. Essentially, changes in the location of upper-level troughs and ridges, whether they are tropical or extratropical in origin, modulate rainfall over the subcontinent. Diab *et al.* (1991) identified eight synoptic types over Natal and found that four of them are responsible for 81% of the precipitation. They used discriminant analysis to prove that the synoptic types are statistically distinct (see Chapter 4). Moving to North Queensland, Australia, Sumner and Bonnell (1986) studied the relationship between wet-season atmospheric circulation and daily rainfall. In this tropical region, the interaction between the air flow associated with rain-producing systems and local topography is the major determinant of rainfall presence and quantity.

Another well-studied region using the circulation-to-environment unique manual approach is the Mediterranean. Prezerakos (1985) compiled a classification of northwest African depressions affecting Greek weather. These depressions do not bring much precipitation to Greece but do produce southerly winds and increased temperatures. Dayan (1986) showed that synoptic conditions over Israel relate to different air-flow trajectories and, therefore, varying levels of airborne pollutants. To show the synoptic-climatological basis of Greek seasons, Maheras (1988, 1989) successfully delimited the winter and summer by applying principal-component analysis to the frequencies of 16 manually-derived weather types.

Finally, circulation-to-environment unique manual schemes have been employed to assess the climate of various regions of North America. In the desert southwest, tropical cyclones off the eastern North Pacific (Douglas and Fritts, 1972) and cut-off lows (Douglas, (1974) have a large impact on precipitation variability. For the midwest, Westerman and Oliver (1985) discovered that interannual variations in the frequencies of six global-scale, upper-air configurations produced the radical winter-to-winter fluctuations experienced throughout the 1960s and 1970s. Over the southern Great Plains in spring, the elevated mixed-layer inversion has well-defined physical structure and temporal evolution (Lanicci and Warner, 1991a). This phenomenon goes through a characteristic life cycle (Lanicci and Warner, 1992b), producing a distinct spatial and temporal pattern in severe storms over Kansas, Oklahoma and Texas (Lanicci and Warner, 1991c).

Environment-to-circulation classifications. Investigators concerned with precipitation often rely on this unique manual approach. This is probably because precipitation varies dramatically in time and space, and the fine tuning provided by environment-to-circulation classifications enables the investigator to focus on only those precipitation (and therefore synoptic) events of interest (see Chapter 6). For example, Prezerakos and Angouridakis (1984) used this approach to discover two blocking patterns over Europe that produce snowfall over Athens. In another application of this strategy, Alijani and Harman (1985) uncovered the uplift mechanisms that generate precipitation events in Iran. Of the 12 examples of environment-to-circulation unique manual classifications mentioned here, 10 deal with precipitation and moisture.

This approach is useful in deducing synoptic controls on droughts and floods. For example, Harman and Harrington (1978) showed that contrasting flood and drought conditions in upper midwestern North America relate to startling contrasts in atmospheric circulation, frontal patterns, and dew-point temperatures. Dey (1982) studied drought in the Canadian Prairies. He found that a quasi-stationary mid-tropospheric ridge over western Canada blocked the normal westerly progression of storm systems, shifting them far north of their normal position and causing stable, dry conditions under the ridge. Hirschboeck (1987a) defined eight hydroclimatic (primarily synoptic-climatological) categories producing flood events in the Gila River Basin, Arizona. From this she demonstrated that floods in the basin originate from several atmospheric circulation types. Importantly, floods are generated in some areas of the basin by different synoptic mechanisms. For instance, frontal activity produces much of the flooding experienced in the northern parts of the basin, while widespread southwest "monsoonal" precipitation generates most of the floods in the south. She also showed that the greatest flooding in the Salt River basin, located to the Gila River's north, results from frontal storms. In a different study, Hirschboeck (1987b) created a synoptic classification for varying scales of catastrophic flood events over North America. She established that all scales of catastrophic flooding can be linked to anomalous circulation patterns. But, she cautions that such anomalous patterns will not always produce catastrophic floods, and that, with flash floods, the flooding response to the circulation anomaly may be extremely localized.

A potpourri of other moisture and precipitation studies can be found in the environment-to-circulation unique manual literature. Harrington and Harman (1985) identified and summarized the synoptic patterns associated with gradients of warm-season moisture stress across the western Great Lakes region of the United States. They showed that surface weather patterns, in association with the mean global-scale flow across North America, produce a distribution of moisture stress consistent with the phytogeography of the region; that is, the prairie-forest ecotone. Harrington and Brown (1985) also used this synoptic approach to explain the bimodal warm-season precipitation distributions characteristic of the upper midwest. They suggest that the mid·summer depression in the precipitation profile is the expected result of seasonal changes in tropospheric circulation patterns. Carleton (1987) established that summer rainfall in the American southwest is related to key anticyclonic synoptic types at upper levels. Changes in the position of the mid-tropospheric ridge over time explain the anomalously wet summers of the 1950s and dry summers of the 1970s. Zangvil and Druian (1990) concocted an environment-to-circulation scheme to see if rainfall over Israel is strongly affected by the trajectories of air before a rain event. They discovered that because of the geography of the Israeli coastline, the effect of air-stream orientation differs across the country: troughs tilting from northeast to southwest produce more rain in southern Israel, while those with northwest-to-southeast tilts produce more in central and northern regions.

Only two notable environment-to-atmosphere unique manual classifications do not deal with precipitation and moisture. Chung (1978) used this approach to study synoptic conditions over southern Canada that favor long-range transport of sulfur dioxide and its oxidation to the acidic pollutant, sulfate. By extracting synoptic patterns associated with high pollution from the data set, Chung found that three configurations are associated with sharp fluctuations of the ambient sulfates: northerly flow behind the passage of a cold front produces low sulfate values; back-of-high situations generate elevated sulfates; and stationary fronts have low sulfate values north of the boundary and high concentrations to its south. Brazel and Nickling (1986) classified the atmospheric circulation according to its association with dust storms over Arizona. They identified four important dust-weather types: frontal; local convective; tropical disturbance; and cut-off low. Although they could trace dust storms to specific synoptic events, they demonstrated that these storms cannot occur unless surface conditions are right: for example, the soil must be dry, loose, and relatively free of vegetation.

Worked example

There are two reasons for presenting a worked example. The first is to demonstrate the construction of a manual classification and all of its attendant problems. The second is to create a classification to use with the environmental scenarios of Chapter 6. In Chapters 3, 4 and 5, I offer worked examples of several other synoptic-classification techniques for the same reasons.

Procedure description

The classic mid-latitude cyclone model forms the basis of the manual synoptic classification. Other authors (for example, Barchet and Davis, 1984) have applied similar generic schemes over North America; such classifications are appropriate for use over western Pennsylvania (Yarnal and Leathers, 1988; Yarnal, 1989), the focus of the environmental scenarios. A slight modification was made to the generic model to accommodate an important local climatic consideration, lake-effect precipitation. A team of five investigators subjectively classified ten years of United States' daily surface weather maps (NOAA, 1978–87) into eight synoptic types.

The training procedure was crucial to the success of the classification. First, each of the five investigators independently classified the first year's daily weather maps into one of the eight categories. The full team then discussed each day's map for that year, comparing individual classification results and reaching a consensus categorization. After completing the first year in this manner, the team "wrapped around" and started classifying the first year's weather maps again. Thus, the training procedure allowed group interpretations to evolve through discussion and experience. When their interpretations on the second pass through the data invariably matched the first-pass categorizations, consistent implementation was assumed and the team jumped to the second year's weather maps. From that point, because of the difficulty in bringing five interpreters together for large blocks of time, classification sessions varied from three to five team members.

Later, two other analysts trained in the same way and compared their results to the first team's classification. Disappointingly, interpretations matched only about 75% of the time. The 25% of the weather maps with different between-group classifications turned out to be the cases that generated the most within-group discussion. In other words, these were maps that were not easy to classify. For the most part, these pressure patterns were transitional synoptic types.

A schematic diagram of the manual typing scheme is presented in Figure 2.1; typical Unites States surface weather maps associated with each of these types is shown in Figure 2.2. Note that these typical maps were selected subjectively *before* the mean sea-level pressure patterns of Figure 2.3 were calculated. The following summarizes each of the eight types and comes from Comrie and Yarnal (1992).

RC   denotes *cyclonic* conditions with *rain* or other precipitation. Skies are cloudy with cyclonic or frontal uplift near warm, stationary or occluded fronts. The wind is typically from the southwest; storms are usually westerly or tracking up the East Coast.

CF   stands for *cold front* passage. CF is associated with the westerly migration of a cyclone to the north of western Pennsylvania and a trailing polar or arctic anticyclone. As the cold front passes and the warm-sector air mass gives way to a cold-sector air mass, temperature drops 5° or more, wind shifts from south to west, and dew-point temperature decreases. This type is usually cloudy with precipitation.

PH$_w$ and **PH$_d$** are associated with *pre-high* anticyclonic conditions; subscripts refer to *wet* and *dry* subtypes to distinguish the presence of lake-effect precipitation. These types typically occur after the passage of a cold front and therefore are associated with rising

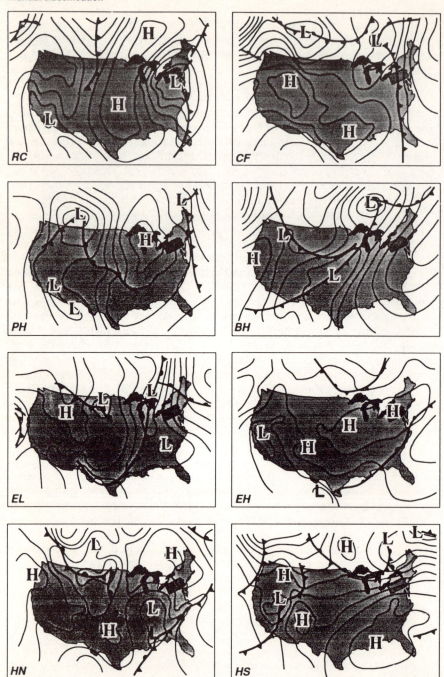

Figure 2.2 Typical surface weather maps associated with the manually derived synoptic types.

pressure and clearing sky. PH_w and PH_d patterns denote either continental air masses from the North American interior or modified maritime-polar air from the North Pacific Ocean; both air-mass types bring a northwest wind component and relatively low dew-point temperature.

BH *(back of high)* is the western side of a high-pressure cell, in which modified tropical air is advected from the south-southwest. This type precedes a cold-front passage and possesses a relatively high dew-point temperature, increasing cloudiness and falling pressure.

HS indicates a *high-pressure cell to the south* and a corresponding low-pressure cell to the north. A component of the wind is from the west. This weather pattern frequently has indeterminate curvature of the isobars and variable cloudiness.

HN is a *high-pressure cell to the north* with a corresponding low-pressure cell to the south. A component of the wind is from the east. HN typically has indeterminate curvature of the isobars and variable cloudiness.

EL stands for *elongated low*. This weather type is a surface trough or depression, sometimes with multiple centers and often overlain by an upper-level trough. It has no frontal boundaries, weak pressure gradients and light wind. EL days are usually cloudy with some precipitation.

EH is an *extended high*: an anticyclone or ridge which extends over a large area, sometimes with multiple centers, and usually in conjunction with an upper-level ridge. It has weak pressure gradients, light and variable wind, and fair weather.

The first four synoptic types (RC, CF, PH, and BH) are closely associated with the classic cyclone model (Figure 2.1, top). The investigators used the other four synoptic classes (HN, HS, EL and EH) when the classic model was inappropriate (see the four bottom panels of Figure 2.1). The latter types usually represent situations with weak pressure gradients, no apparent surface fronts, and north–south rather than east–west configurations.

The classification process involved studying the various weather-map elements in relation to western Pennsylvania; note Pennsylvania's position in Figure 2.2. These elements include surface-pressure patterns, fronts, areas of precipitation, and individual station data. The Pittsburgh weather station (designated *P* on Figure 2.3 and similar figures in Chapters 3, 4 and 5) operated throughout the period and guided much of the interpretation.

One problem with static classifications is that they reflect the weather at one instant; in this case 1200GMT (0700 local time). That instant might not represent the weather for most of the day. Picking a hypothetical day, the Pittsburgh station model might show that at 7:00 a.m. conditions were

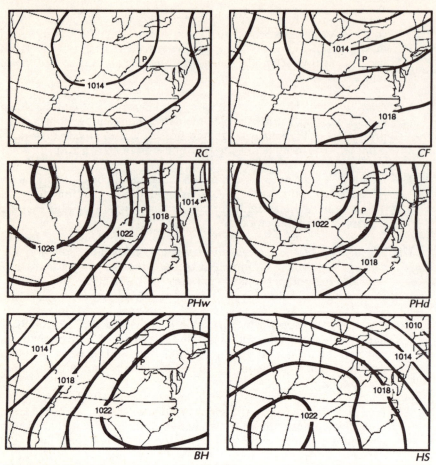

Figure 2.3 Manual map-pattern classification: average sea-level pressure pattern (millibars) associated with the synoptic types.

warm, humid, and cloudy, with a southwesterly wind — typical BH weather. However, later that morning, a strong cold front (CF) passed the station, with heavy cloud, violent weather, and 25 mm of precipitation in its wake. By noon, the front was replaced by a continental polar air mass, with cool, dry northwesterly flow, a relatively low dew-point temperature, and rapidly clearing sky. The remainder of the day maintained these PH weather characteristics.

What synoptic type is this hypothetical day? Arguments can be made for three possibilities. At the synoptic hour, western Pennsylvania was under a back-of-high regime and probably had been for some time. Some would argue that the weather at the synoptic hour should be the only guide to the analyst; thus the synoptic type had to be BH. Others would argue that the cold-front passage was the most important weather development of the day, with the most impact on the surface environment. If the front spawned a

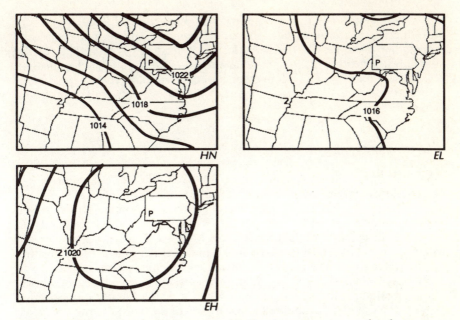

Figure 2.3 cont. Manual map-pattern classification: average sea-level pressure pattern (millibars) associated with the synoptic types.

tornado that hit downtown Pittsburgh, they would find it absurd to categorize the day anything but CF. The final argument favors PH since PH conditions dominated one-half of the day.

It is clear that classifying the day as either BH, CF, or PH is problematic. Classifying it BH does not account for the cold front and its attendant weather or the trailing polar air mass, whereas classifying the day CF would ignore the roughly 20 hours that were not CF. Finally, classifying the day PH disregards the cold front and the strong precursor (BH) conditions. Furthermore, suppose the previous day started with 15 hours of PH conditions, but finished with nine of BH. Over 16 hours of strong advection of maritime tropical air would never show up in the climatological record. In this hypothetical case, the investigator must choose one of the three synoptic classes, but each will induce considerable variance in the data.

To take account of this problem, the five analysts added a "kinematic" component to their interpretations. They studied the weather maps of the previous and following day; they observed the pattern of cloud, precipitation, temperature, dew-point temperature, pressure tendency, and wind speed and direction over western Pennsylvania on the map day and on the previous and following days; and they estimated the synoptic type that dominated the day's weather in terms of the proportion of the day it was present and of its surface environmental impacts. From this appraisal, they chose the synoptic type that best captured the general character of the day's weather. This technique could be refined to focus on one specific application, but this would mean that classification might be inappropriate for other environmental problems. Although the strategy used here introduces even more subjectivity to the analysis and makes the results impossible to duplicate, the investigators considered it a strength of the manual classification technique. Of all the automated synoptic-classification schemes, only the eigenvector-based synoptic-typing procedure approximates this holistic assessment of the day's weather over time (see Chapter 4).

Classification results

The analysts classified over 95% of the study period's 3,652 days into one of the eight types; the remaining days were either unclassifiable (**OT**: *other*) or had missing weather maps. The classification was dominated by frontal types. For the ten-year study period, RC was the most frequent type, occurring nearly 24% of the time (Figure 2.4, top). This is not surprising for two reasons: (1) western Pennsylvania is under or near the average position of the polar front; and (2) the RC category includes warm, occluded and stationary fronts, while the CF category has just one frontal type. Therefore, the high numbers for RC are partly a function of the classification scheme. Together RC and CF accounted for over 36% of the period's synoptic types. In comparison, the main anticyclonic frontal types, BH, PH_w and PH_d, existed on 15%, 5% and 9% of all days, respectively, for a total of 29%. Thus, despite the region's intimate relationship to the polar front and the classification's aim to take advantage of that fact, the classic frontal model accounted for only two-thirds of western Pennsylvania's weather patterns during the study period.

Because one-third of the days had weather patterns that could not be explained by the frontal model, it was important to have included variants of the model in the classification scheme. This is especially true for EH, the second-most-frequent synoptic type, which appeared nearly 18% of the time. Large upper-air ridges frequently occur over the eastern United States, with no noticeable surface fronts in the vicinity. Upper-air troughs with little surface expression (EL) are also important to climate, occurring on more than 6% of all days, more than the region's famed lake-effect precipitation (PH_w; see later in this section). The north–south surface configurations HS and HN occurred 5% and 2% of the time, respectively.

The OT category must not be neglected. Although these unclassifiable maps accounted for only about 1% of the study period, significant weather events can occur on these days. Most important are the tropical storms which occasionally invade the area. For example, in one day, 1972's Hurricane Agnes brought 150–300 mm of rainfall to western and central Pennsylvania. Hurricane David (1979), and several other tropical systems that invaded the area within the study period, brought far less rainfall than Agnes, but still had significant impact on the region's environment.

Mean pressure maps, constructed by averaging surface-pressure grids for all days with a given synoptic type, differ slightly from the typical weather patterns selected by the analysts before the classification (Figures 2.3 and

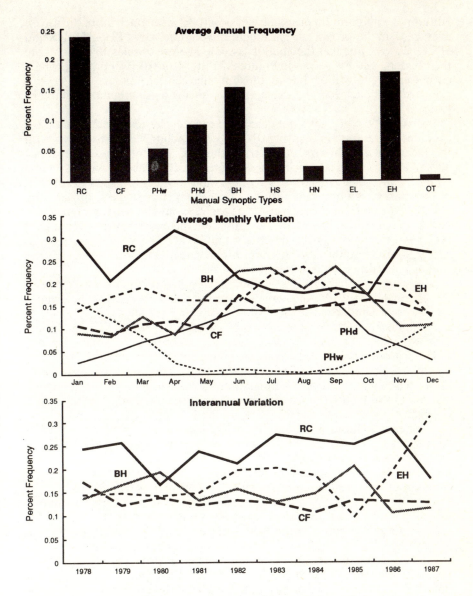

Figure 2.4 Average annual frequency of the manually-derived synoptic types (top), average monthly variation of six types (middle), and interannual variation of the most frequent low- and high-pressure types (bottom).

2.2, respectively). For instance, cold fronts associated with CF and RC types were envisaged to have a north–south orientation, while in reality they tended to have a distinct west–southwest orientation. In contrast, EH had a strong meridional orientation, as opposed to the more amorphous latitudinal configuration originally supposed. EL was much more diffuse on average than the investigators thought it would be. Only PH, BH, HS and HN types were much as the analysts assumed before the classification. Note that the PH_w and PH_d-types in Figure 2.3 differ from the wet pattern, which occurs mainly in winter, having much higher central pressure and steeper pressure gradients than the dry pattern, which is predominantly a summer pattern.

The synoptic types have noticeable seasonal cycles (Figure 2.4, middle). From late autumn to mid-spring, RC occurred about 30% of the time, with the exception of February. In contrast, this type was only present on about 20% of all days from late spring to mid-autumn. The other principal cyclonic type, CF, was observed about 10% of the time in winter and spring, but 15% of the time in summer and autumn. These two facts suggest, in part, the annual expansion and contraction of the circumpolar vortex. When the vortex expands southward in late autumn, the more northerly parts of cyclones tend to pass over the study area; when the vortex contracts in spring, more often the low-pressure centers pass to the north, with only the trailing cold front affecting western Pennsylvania.

A distinct seasonal cycle in the main high-pressure types is also evident. BH averaged about 10% in winter and spring, but over twice that in summer and autumn. Similarly, roughly 5% more upper-level ridges (EH) occurred in summer and autumn than in winter and spring. Again, these numbers mirror the annual movement of the circumpolar vortex.

The seasonal cycle of PH_w relates to lake-effect precipitation. Lake-effect precipitation, chiefly in the form of snow, occurs when northwesterly winds blow continental-polar or continental-arctic air masses over the relatively warmer Great Lakes. The lake surface warms the overlying cold air. More important, the steep vapor-pressure gradient between the dry air and lake causes heavy evaporation, humidifying the lower layers of the atmosphere. This warmed, humidified, near-surface air in now unstable and rises. As it ascends, it cools, condenses, and releases latent heat, which makes the air mass even more buoyant. This process forms a heavy deck of stratocumulus clouds over the Great Lakes that extends a few hundred kilometers beyond the lee shore. Worse, frictional drag, caused when the air mass encounters the rougher land surface, piles up these clouds and releases sometimes massive amounts of precipitation along the lee shore. Because many of these areas are backed by highlands, snowfall totals can be significant both along the lake shore and in the mountains. This is primarily a wintertime phenomenon and, even then, can only exist as long as the lakes are unfrozen. In Figure 2.4 (middle), note that PH_w reaches near-zero values for the warmer months, but occurs on more than 15% of the mid-winter days.

The annual PH_d cycle is nearly a mirror image of PH_w. In summer, the Great Lakes are cooler than the northwesterly airflows associated with post-frontal, pre-high conditions. The lake now chills the base of the air stream, stabilizing it. Clouds tend not to form and there is no precipitation. The combined total of PH_w and PH_d days (not shown) has two peak seasons, summer and winter, accounting for roughly 16% of the days. Percentages in the transitional seasons hover around 12%. Because there is no reason to think that the seasonal cycle of post-frontal high pressure should have this double-peak structure (indeed, BH and EH were more common in summer than winter), this suggests that the interpreters tended to bias their classification toward unambiguous situations where the lake-effect was clearly one way or the other.

Interannual variation in synoptic-type frequencies is also present in the data (Figure 2.4, bottom). The most variable of the study period's major types proved to be EH, with a range of over 22%. Its lowest frequency was 9%, while its high was over 31%; nearly one-third of the days in 1987. In contrast, the annual total of cold fronts (about 14%) was amazingly stable throughout the study period. RC and BH are strongly anti-correlated, and both fluctuate within similarly large ranges: 12% for RC and 10% for BH.

3 Correlation-based map-pattern classification

The difficulties of manual map-pattern classifications discussed in Chapter 2 prompted many synoptic climatologists to embrace the computer-based techniques that proliferated in the 1970s and 1980s. Not only did practitioners hail these new classification methods as easier to use, but they also assumed them to be more objective (Barry, 1980). With several years of experience using automated classification procedures, synoptic climatologists now realize that each of these techniques involves a number of subjective decisions. Furthermore, it is not clear whether computer-based synoptic-climatological classifications are an improvement on manual techniques in their ability to relate the atmospheric circulation to the surface climate. Nevertheless, automated schemes are quick and easy to use and investigators can replicate the results, two things that cannot be said for manual classifications.

Correlation-based map-pattern classification is one of the two most important categories of automated synoptic climatology. There have been few changes to the procedure since Lund (1963) introduced it. The reasoning behind it is straightforward: the machine executes the same task the manual analyst performs, placing similar map patterns into discrete categories. Perhaps the correlation-based technique's popularity springs from this intuitive, simple basis. The fact that the input to and output from the computer is a real, readily interpretable pressure pattern also adds to the procedure's attraction. This contrasts sharply with the other main category of computer-assisted classification, the eigenvector techniques to be covered in Chapter 4. The results of an eigenvector analysis are mathematically compact but are often difficult to interpret and lack the uncomplicated appeal of a weather map.

Correlation-based studies take the circulation-to-environment approach to synoptic climatology (Figure 1.1). The atmospheric circulation data are classified before the environmental data are addressed. Thus, the circulation categories are independent of the surface. This is important because it confirms that correlation-based synoptic classifications are suitable for use with model-comparison statistics (Chapter 6).

The correlation-based classification technique is the subject of this chapter. First, I address the classification methodology, discussing the basic procedure and subjective decisions facing the investigator. Then I review the literature, showing examples of this classification scheme and its application to environmental problems. Finally, I present a worked example of the correlation-based classification technique.

Methodology

Basic procedure

There are two correlation-based techniques for the classification of weather maps. The first, introduced by Lund (1963), uses Pearson product-moment correlations (r_{xy}) to establish the degree of similarity between map pairs:

$$r_{xy} = \frac{\displaystyle\sum_{i=1}^{N} \left[(x_i - \bar{X})(y_i - \bar{Y}) \right]}{\displaystyle\left[\sum_{i=1}^{N} (x_i - \bar{X})^2 \sum_{i=1}^{N} (y_i - \bar{Y})^2 \right.} \tag{3.1}$$

In this formula, x_i represents the variable (almost always standardized pressure) at each of the first map's N points, while y_i is the same variable at the same points on the second map. $\bar{X}$ and $\bar{Y}$ represent the means of the N-point grids. Kirchhofer (1973) put forward the second technique, (3.2), which uses a sums-of-squares algorithm to compute the correspondence between weather maps.

$$S = \sum_{i=1}^{N} (Z_{xi} - Z_{yi})^2 \tag{3.2}$$

Here, S is the Kirchhofer score, Z_{xi} is the normalized grid value of point i on day x, Z_{yi} is the normalized grid value of point i on day y, and N is the number of data points. Yarnal (1984b) presented a computer program based on the Kirchhofer technique which, as shown below, is appropriate to use for all correlation-based classifications.

Through a simple mathematical transformation, Willmott (1987) demonstrated that the Lund correlation coefficient and Kirchhofer metric are one and the same and can be related to one another by the formula:

$$S = 2(1 - r) \tag{3.3}$$

Because Lund preceded Kirchhofer, Willmott suggested that the technique should somehow be attributed to Lund. Yarnal and White (1987) agreed, recommending that all such techniques be called *correlation-based*. In the remainder of this book, I treat the two procedures as identical and do not differentiate between them. Furthermore, for comparability, I convert all Kirchhofer scores from the literature into the more-familiar correlation coefficients.

The flow chart in Figure 3.1 shows the correlation-based classification methodology. In the remainder of this section, I describe the sequence of activities displayed in the rectangular boxes; these are the *basic steps* of the

49

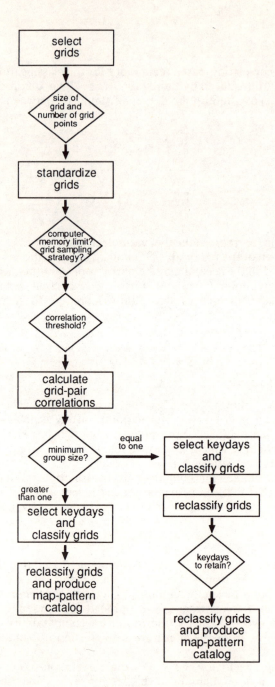

Figure 3.1 Correlation-based classification of map patterns.

procedure. Later in the chapter, I discuss the questions appearing in the diamond-shaped boxes; these are the required, subjective *investigator decisions*.

The first step of the procedure is to obtain the map-pattern data to be classified (Figure 3.1: *SELECT GRIDS*). The widespread availability of large, once- or twice-daily gridded data sets have made their use the norm in automated classification. Virtually all correlation-based classifications employ mean sea-level pressure or pressure-surface data. Therefore, the following description assumes the data are pressure grids.

Because synoptic climatologies often span more than one season of the year, it is customary to standardize the grids before classification (Figure 3.1: *STANDARDIZE GRIDS*). Standardization removes the seasonal influence on absolute pressure or geopotential height. For instance, because of the warmer and therefore thicker summertime troposphere, summer geopotential heights are much greater than those of winter. In addition, standardization eliminates the seasonal impact on pressure-pattern intensity (for example, wintertime pressure gradients are much steeper than those of summer). Therefore, with the seasonal effects on absolute pressure and pressure gradient removed, only the generalized map pattern remains, and pressure configurations of various months and seasons are comparable. The common Z-transformation is used:

$$Z_i = \frac{x_i - \overline{X}}{s} \tag{3.4}$$

where Z_i represents the standardized value of grid-point i, x_i is the observed value at grid point i, $\overline{X}$ is the mean of the N-point grid, and s is the standard deviation of the grid. The daily value of $\overline{X}$ can be retained as an index of mean pressure or geopotential height, and daily values of s can be used as a measure of pressure-pattern intensity across the grid (Yarnal, 1985a).

Each standardized grid is then compared to all other grids (Figure 3.1: *CALCULATE GRID-PAIR CORRELATIONS*). The correlation coefficient calculated for each pair denotes overall pattern similarity. Prior to this step, the investigator must choose a correlation-coefficient threshold; I discuss the critical ramifications of this decision later in the chapter. Average values for r typically range from 0.5 to 0.7. Grid-pairs that reach this cut-off are "significant," and the program stores the correlation coefficient for subsequent use. The program assigns a "not significant" descriptor to those correlations that do not pass this threshold, storing these in the computer, too.

Although the overall correlation between a pair of grids may be high, specific zones of the maps may have widely differing patterns. Therefore, to insure pattern similarity in all areas of the grids, many versions of the correlation-based procedure check for correspondence between map sectors. For a pair of rectangular grids, this is simply a matter of calculating a correlation coefficient between all corresponding rows and columns. Those map patterns that have overall statistical similarity but fail this sectoral test are labeled "not significant."

Once the correlation coefficients have been calculated for each pair of grids, the procedure selects the map-pattern categories (Figure 3.1: *SELECT KEYDAYS AND CLASSIFY GRIDS*). This involves an iterative process in which the computer scans the previously calculated correlations for that grid with the greatest number of "significant" correlations. The program recognizes this as the most typical grid pattern and calls it Keyday 1. The algorithm removes that grid and each grid that is "significantly" correlated with it. The program then repeats the procedure, identifying Keyday 2 and removing all grids "significantly" associated with it. The program continues to identify keydays and remove correlated map patterns until reaching an investigator-specified *group-size minimum* for the map-pattern classes. Typical group-size minima range from five to 50 grids. If the investigator sets this number at one (that is, the classification includes unique map patterns), the flow chart follows an alternate route. I discuss this path, the problem with specifying a group-size minimum, and the nature of the decision facing the investigator later in the chapter.

After selecting the keydays, each grid must be reclassified (Figure 3.1: *RECLASSIFY GRIDS AND PRODUCE MAP-PATTERN CATALOG*). This is necessary because it is possible for any grid to be "significantly" correlated with more than one grid. In other words, a grid may have been misclassified during the keyday selection procedure. For example, given a threshold of $r = 0.7$, a grid may have had a correlation with Keyday 1 of 0.71 and therefore was removed with it in the process described in the previous paragraph. However, that grid may have had a correlation with Keyday 6 of 0.93, so the correct classification should have been map-pattern 6. In the reclassification procedure, the algorithm calculates the correlation between each grid and each of the m keydays; it also determines row and column sectoral correlations for each of the keydays. At this point in the procedure, each grid is assigned to a map-pattern category based on its highest significant correlation. Those days that have no significant correlations with the keydays or that have widely varying sectoral correlations are "unclassified." It is typical for class memberships to change noticeably at this stage, with the low-order map patterns (1, 2 and 3) usually losing members and higher-order map patterns gaining members. The output from this step is a map-pattern catalog stored in spreadsheet form for statistical and graphical analysis.

The search for synoptic-climatological relationships with surface environmental variables can now begin. Before reviewing previous applications of the correlation-based classifications to surface variables, however, I discuss the choices facing the investigator and the problems with this classification procedure.

Investigator decisions

The correlation-based classification procedure requires the investigator to make a number of subjective decisions. These choices determine the following: (1) number of map patterns; (2) within-group similarity and between-group differentiation; and (3) percentage of grids classified. Unfortunately, these decisions are not independent but are inextricably intertwined. For example, to classify a large percentage of the grids, the number of map patterns must increase or within-group variation must increase, while between-group differentiation decreases. If a small, manageable number of map patterns is appropriate, then the percentage of grids classified must decrease or the investigator must relax the statistical borders among the map-pattern categories. Finally, if an analysis requires distinct, well-defined classes, then the percentage of grids classified must decrease or the number of map patterns must increase dramatically. In the final analysis, the balance among these factors should reflect the needs of the application for which the classification is being developed.

Before the classification procedure can begin, the investigator must decide on the size of the grid and the number of points in it (Figure 3.1: *Size of grid and number of grid points?*). Yarnal and White (1987) demonstrated that the greater the number of grid points, the more variety that is possible in the grids, and the more map patterns that will result. Similarly, with grid-point spacing held constant, larger areas will produce more map patterns. The investigator can reduce the number of map patterns by decreasing the size of the study area or by increasing the spacing of the grid points (that is, reducing the number of grid points). Unfortunately, there is a danger in increasing the distance between grid-points: smaller synoptic features can be filtered, so that only long waves remain (Yarnal, 1984a). The impact of this on the results can be disastrous if the climate of a region is dependent on high-wavenumber, high-frequency synoptic events. Likewise, shrinking the size of the gridded area can have considerable impact on the results if the area is too small to resolve the circulation features affecting a region's climate.

The algorithm conventionally used to calculate grid-pair correlations demands huge amounts of central memory in a computer (Figure 3.1: *Computer memory limit?*). The memory required is calculated:

$$RAM = (n\,(n-1))/2 \tag{3.5}$$

where n is the number of grids in the sample and RAM is the amount of random-access memory needed to store the grid-pair correlations (Yarnal, 1984b; Yarnal and White, 1987). RAM grows exponentially as the sample size increases. The smaller, slower computers of the 1970s to mid-1980s effectively restricted all investigations to no more than five years. Even today, computer resources limit the number of grid-pairs classified with this algorithm. For example, with special permission, the maximum RAM accessible for a correlation-based classification on the current Penn State mainframe is 16 megabytes, which, using formula 3.5, translates into just over 15 years' daily grids. A 30-year classification of daily grids requires 60 megabytes of RAM! Therefore, unless investigators have access to a supercomputer, analyses for periods longer than a decade or so call for an alternate algorithm.

Investigators have developed a number of strategies to handle this restriction on the number of grids used to determine the keydays (Figure 3.1: *Grid sampling strategy*?). For instance, Petzold (1982) limited the duration of his study to just three lake-ice thaw-to-freeze seasons; Bradley and England (1979) took a random sample (2,128 days) of their 29-year (10,591-day) period; Moritz (1979) chose a block of the most recent five years to represent his 29-year study period; and Yarnal (1984c) selected four specific years to represent the climatic mean and extremes of his study period. Despite the apparent success of these strategies, it is highly desirable to utilize the entire population of grids. Yarnal and White (1987) have demonstrated that the results of a correlation-based classification can vary with differing samples, while Yarnal *et al*. (1988) have shown that subsequent synoptic-climatological relationships with surface variables may change with varying classifications. Another problem with using a sample of grids is that the map patterns passing over a region may vary with time. Thus, the schemes of Moritz (1979) and Yarnal (1984c) given above may be inappropriate for long-period synoptic classification.

A conceptual model of an improved correlation-based classification algorithm utilizes a simple iterative procedure to calculate correlation coefficients and compare grids simultaneously. To be viable, this strategy must produce results that are identical to those of the conventional procedure. The chief difference between the two algorithms is that the new one would be computationally intensive, requiring very little computer memory but large amounts of computer time. In contrast, the standard procedure is RAM-intensive. Using the new algorithm, a correlation-based classification could run on any computer that has sufficient memory to store the grids. A mainframe or mini-computer could manage a correlation-based classification of virtually any length. In fact, many micro-computers could handle the task, although they might take hours to days to run the classification.

For the classification procedure to commence, the investigator must specify a correlation threshold (Figure 3.1: *Correlation threshold*?). Several investigators have focused on the effect of varying threshold limits on the results of a correlation-based synoptic climatology (Sabin, 1974; Moritz, 1979; Petzold, 1982, Hoard and Lee, 1986; Key and Crane, 1986). Their findings all show that with higher thresholds, within-group variance decreases and between-group differences increase; with lower threshold values, the differences between map patterns blur and the internal variation of each map pattern increases. One of the primary goals of classification is to maximize between-group distance and minimize within-group variance, so it would seem that using high thresholds is desirable.

There are two problems with setting high threshold values. First, the higher the threshold, the greater the number of map patterns. It is not unusual to discover that thresholds of $r = 0.8$ or 0.9 produce several dozen map patterns. Most investigators would find this many categories unmanageable; thus, they must consider the option of relaxing the threshold values. Another option is to maintain high thresholds, but to lump subjectively the map patterns into a smaller number of more general, related groups. This idea would have broad appeal if it were not for the second problem with high thresholds: the percentage of grids classified decreases as the threshold rises. A threshold of $r = 0.9$, for instance, might produce classifications with several dozen map patterns, but only half of the grids might be classified. This is also usually unacceptable. Therefore, most scientists opt for relaxing the thresholds to produce a workable number of map-pattern categories and to increase the number of classified grids.

Relaxing correlation thresholds produces more internal diversity within, and less separation among map patterns, so the critical question is: how much messiness within and between map patterns is the investigator willing to accept? Responses to this question vary with research needs and temperament. Some scientists have research designs requiring a small number of classes or a large percentage of classified grids. In contrast, other investigators may have objectives in which the number of classes makes little difference. Also, some researchers are more tolerant of ambiguity and are accepting of less-compact clusters of synoptic categories. Others are less tolerant of uncertainty and less willing to agree to sloppy synoptic categories; they prefer a large number of map patterns that capture every conceivable synoptic situation. Low thresholds used by the more-tolerant investigators can be as low as $r = 0.3$ or 0.4, while those employed by the more-fastidious scientists will approach $r = 0.9$. Typical compromise values range from $r = 0.5$ to 0.7.

The above discussion has focused on the general map-pattern correlation coefficient but can be extended to sectoral r-values. Holding the overall r constant, higher thresholds for sectoral correlations produce more map patterns and fewer classified grids; low sectoral correlation thresholds produce fewer map patterns and a higher percentage of classified grids. The balance between the overall correlation coefficient and the sectoral r is another subjective decision that the investigator must make.

To set a cut-off for the smallest acceptable membership in a map-pattern category, most investigators place a minimum group-size algorithm in the keyday-determination program (Figure 3.1: *Minimum group size? greater than one*). There are two rules of thumb commonly used to define this lower limit. In one, the investigator arbitrarily chooses a percentage of the total number of grids. One per cent is a typical figure so that, in a sample of 500 grids, a group-size minimum of five is appropriate. Another idea is to use one grid per year of daily grids. In this case, if grids from 10 years of data are being classified, the investigator would employ a minimum group size of 10. Changing the minimum group size does not change the keydays selected by the procedure (Key and Crane, 1986).

Petzold (1982) concluded, however, that the use of a minimum group-size algorithm may miss important map patterns. Because the keyday-selection procedure removes their correlates, important synoptic categories can fall below the minimum group-size threshold. If all grids are classified (that is, the group-size minimum is one), the reclassification procedure will determine whether any important map patterns have been missed or not. The investigator can apply a group-size minimum *after* reclassification.

Later in the chapter, I present the results of a correlation-based classification using a group-size minimum of one (Figure 3.1: *Minimum group size? equal to one*). A large number of one-member synoptic groups result from the keyday selection routine. The reclassification procedure found that a number of grids with too few members to have passed even relatively loose group-size minimum requirements (for example, a group-size minimum of 1% of the study period's grids) were, in fact, important map patterns. To set a group-size minimum after reclassification (Figure 3.1: *Keydays to retain?*), a scree plot was constructed (Figure 3.2). On such a plot, the percentage of grids classified, the number of map patterns, and the natural breaks in the distribution are evident. After choosing an appropriate number of map patterns from the scree plot, the investigator again reclassifies the daily grids to create the map-pattern catalog.

In summary, an investigator using correlation-based synoptic climatology must make a number of subjective decisions, all of which will directly influence the results. In the next section, I review the many applications of the correlation-based methodology to surface environmental conditions.

Previous studies

Since its introduction by Lund (1963), the correlation-based classification method has been applied in diverse ways to many different surface environments. A few of these studies focused on the method itself, while others spotlighted the application. Still, most authors paid careful attention to both sides of the synoptic-climatological coin. In the following, I briefly review the correlation-based literature. First I address methodological concerns that were not covered in the previous section or that warrant expansion, then I discuss applications of correlation-based map-pattern classification to surface environments.

Methodological studies

In his seminal article, Lund classified five winters' surface-pressure data for 22 unevenly spaced stations in the northeastern United States. He first set out the correlation-based typing procedure which has been followed, more or less, to this day. He used $r = 0.7$, although he did try $r = 0.9$ but decided this produced too many unclassified days. He ran the procedure until 10 keydays were identified, stopped there, and categorized all of the days *vis-à-vis* these keydays. Lund related each of the map patterns to precipitation, sunshine, and snowfall classes at Boston. He showed experienced forecasters the 10 keyday maps and asked them to identify the one occurring most often, the one associated with the heaviest snowfall, and the one with heaviest cloud over Boston. They all picked the keydays identified by the automated synoptic climatology as most common, snowiest, and cloudiest, respectively.

In an early technical study, Scholefield (1973) tried to understand how the correlation-based classification scheme works. To do this, he created 10

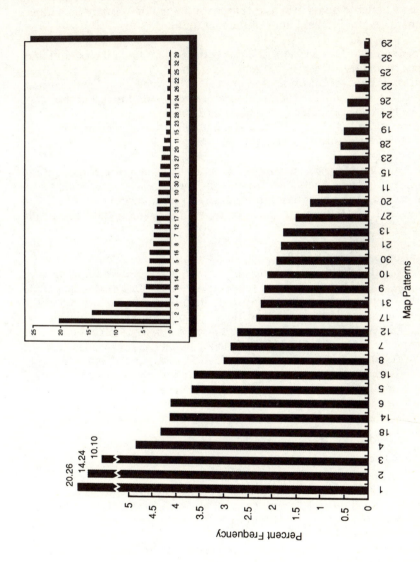

Figure 3.2 Scree-plot used to determine the number of keydays retained in a correlation-based map-pattern classification. Note: one-day map patterns are not shown.

maps with 12 points. Nine maps were hypothetical and one was observed; all had concentric circles around the center of the grid. Holding the isobars drawn on the maps constant, he compared their pressure gradients, the absolute magnitude of their gradients, and their pressure-gradient direction (that is, whether a given pressure center was a high or a low). His findings illuminate several fundamentals of the methodology.

— Correlation values are almost entirely dependent upon the overall map pattern and the pressure-gradient direction.
— Differences in the absolute pressure values between maps have no effect on correlation values when pressure gradients are equal and the maps have the same arrangement of highs and lows.
— Correlations between maps will still be high if the arrangement of highs and lows is roughly the same but pressure gradients are quite different.

Thus, correlation-based classification is at heart a pattern-recognition technique. It focuses on the pattern of highs and lows across the grid, not on the strength of the pressure features. It cannot even discriminate between summer and winter pressure patterns unless the gradients are drastically different.

As Lund (1963) did in his founding investigation, all users of the correlation-based method have to wrestle with the problem of setting a correlation threshold. In a little-known yet reasonably comprehensive study, Sabin (1974) attacked two related problems with thresholds. (1) He sought to optimize the trade-off between map-pattern correlation and the number of map patterns produced. High correlation values generate too many, highly specific map patterns (for example, when $r = 1.00$, the number of map patterns equals the number of maps classified); low r values produce few, overgeneralized map patterns. (2) He wanted to figure out the threshold which produces the strongest relationship between the map patterns and the surface climate. To address these problems, he used 403 January surface-pressure grids with 12 points, adopted a minimum group size of 1% (that is, 4 grids), and ran classifications using $r = 0.7, 0.8$ and 0.9. He found the following facts.

— The $r = 0.7$ and 0.8 computer runs classified 82.2% and 71.9% of the grids, respectively, while the 0.9 run classified 38.8% of the grids.
— The $r = 0.7$ and 0.8 classifications generated map patterns that were nearly equivalent and captured clear high- and low-pressure centers; 0.9 map patterns were quite different and tended to capture transitional phases between high and low regimes.
— The standard deviations around the group means were large for the $r = 0.7$ run but relatively small for the other two runs.
— The relationships between the surface climate and the $r = 0.7$ map patterns were much weaker than those of the 0.8 and 0.9 classifications because the inclusion of less similar maps introduces contamination by extraneous climatological data. There was little difference in the relationships between the surface and the map patterns at $r = 0.8$ and 0.9.

From these results, Sabin concluded that $r = 0.8$ is the "best compromise" threshold. This conclusion deserves discussion but that must wait until the following problems are considered.

Grid selection produces dilemmas for investigators using the correlation-based and, presumably, other automated classification techniques. Overland and Hiester (1980) found that gridded pressure data may be too coarse to resolve some important small-scale atmospheric features. They concluded that if an investigation calls for finer resolution than the gridded data can supply, the correlation-based technique is inappropriate and subjective interpretation of hand-plotted charts or satellite images is the only remedy. Expanding on this line of reasoning, Yarnal (1984a) conducted separate but identical correlation-based classifications on two different 500-mb gridded data sets. Each grid was five-by-six, but one had a resolution of approximately 350 km, while the other had grid spacing of about 700 km. Correlation thresholds and other investigator-controlled parameters were held constant. The results demonstrated that the fine-resolution grid accentuated high-wavenumber, high-frequency synoptic features, while the coarse grid spotlighted planetary-scale long waves. Increasing the grid spacing effectively filtered out small-scale weather phenomena. Thus, investigators working with gridded data must know the scale of the atmospheric-circulation feature they want to study and which grid spacing will emphasize that scale.

Another potential pitfall associated with the correlation-based classification of gridded data is the number of grid points selected by the investigator. Holding grid spacing constant, Yarnal and White (1987) classified 30-, 56-, and 90-point grids; that is, successively larger areas centered on the same location. Their work showed that, all other factors being equal, more grid points result in a lower proportion of classified grids and a larger number of map patterns. These findings are attributable to the increasing detail in the pressure field as more high- and low- pressure features are added. In a subsequent study, Yarnal *et al.* (1988) demonstrated that changing the number of grid points used in a classification alters relationships between the map patterns and surface climate, with more grid points (that is, larger areas) producing slightly higher variances. The findings of these two studies suggest that, to improve results, investigators should try to use the smallest number of grid points (that is, the smallest area) that will enclose the atmospheric features of interest.

In summary, investigators must take seriously the selection of the grid to be used in a correlation-based classification. They must give careful consideration to: (1) the scale of the atmospheric circulation features assumed to control the surface environment; (2) the grid spacing that will emphasize that wavenumber; and (3) the minimum number of grid points needed to resolve features of that scale. If there is uncertainty over these considerations, then it is best to make a number of experimental classification runs, zeroing in on the optimum number of grid points and their spacing.

Now Sabin's belief that $r = 0.8$ is the "best compromise" correlation threshold can be addressed. Investigators should never use as few as 12 grid points because that number will either cover an area too small to resolve all but the smallest synoptic-scale features or the points will be so widely spread that the grid will filter all high-wavenumber features. However, with more grid points, Sabin's "best compromise" would certainly change to a lower correlation threshold than 0.8 as the inter-map correlations decreased. Hence, Sabin's conclusion should only be taken relatively: that is, in any correlation-based analysis, there is a certain correlation threshold that will produce (1) a relatively large proportion of classified days and (2) strong relationships between the map patterns and the surface environment. It is up to the synoptic climatologist to decide how important finding that optimum threshold is to his or her study.

Blasing (1979) developed a technique to circumvent the problem of choosing a correlation threshold. In so doing, he formulated a significant departure from the procedure depicted in Figure 3.1. In Blasing's method, the investigator selects a *level of generality (g)*, which is the number of grids to be averaged for each keyday. After finding the grid that is most highly correlated with all other grids in the data set, the investigator removes and averages that grid and the $g - 1$ grids that are most highly correlated with it. The resulting pattern is Keyday 1. The investigator continues the process until leaving only $g - 1$ or fewer grids. Thus, the investigator does not set a correlation threshold. In fact, this procedure means that if the investigator used a correlation threshold, it would vary from pattern to pattern. The only judgment on the investigator's part is the number of maps for g. In the most specific case, $g - 1$, each map is its own keyday; in the most general case, $g = N$, the average of all N maps produces one map pattern. Hence, g determines how general the investigator wants the keydays to be.

For experimental purposes, Blasing used 77 Northern Hemisphere surface-pressure grids with 180 points each. He chose $g = 5$ and $g = 10$. The more specific $g = 5$ produced 10 patterns, while the more general $g = 10$ generated just four patterns. Focusing on the $g = 10$ classification, many grids had overall r values less than 0.5 when compared to the average map for their map-pattern grouping. Blasing discarded these poorly correlated maps, keeping 44 of the 77 grids.

Building upon Blasing (1975), Blasing (1979) concluded that his scheme produces results that are comparable to eigenvector-based map-pattern classification. He believed this method is better than eigenvector-based techniques; see Chapter 4 for his point-by-point comparison of correlation and eigenvector map-pattern classifications. Blasing did not compare the results from his scheme to those produced by the traditional correlation-based method. He used this technique to identity global-scale climatic complexes over the North Pacific and North America (Blasing and Lofgren, 1980) and over the Northern Hemisphere (Blasing, 1981), but he did not relate them to regional surface environments.

Applications to surface environments

The correlation-based classification method has been used to relate the atmospheric circulation to many different types of surface environment and environmental variable, ranging from general regional climatologies to single-city air-quality studies. For example, Moritz (1979) conducted a general synoptic climatology of the Beaufort Sea Coast of Alaska. He determined that wintertime pressure fields are dominated by the Arctic high to the north and the Aleutian low to the south. There is a reversal of this general pattern in summer, with the expansion of the North Pacific high into the southeast of the study region and lows in an arc extending northward from the west to the northeast. Transitional seasons have no distinctive circulations but are unique blends of the winter and summer patterns. He found that the classification produced weather-element populations with highly significant differences between the map patterns. The best distinctions were for various temperature measures, but wind speeds, wind directions, and precipitation variables also related well to the patterns. The differences were not as clear for sky cover and dew-point temperature in some seasons, although in others they were quite significant.

A number of relatively long-term (at least 25-year) correlation-based climatologies have been developed to address the relationships between synoptic-scale circulation patterns and regional climatic variation. Bradley and England (1979) used this tactic to study secular changes in ablation-season temperatures and annual precipitation totals at three High Arctic stations. The scientists related map-pattern frequency changes to variations in surface climate after establishing the synoptic controls on these variables. They demonstrated that changing frequencies account for some of the observed climatic variability but not all. They concluded that within-group variations must account for the rest of the unexplained climatic shifts. Keen (1980) used the correlation-based method to determine whether changes in the frequency of synoptic-scale systems can explain the interannual and interdecadal climatic variability around Baffin Island. He looked at differences between zonal and meridional periods (see Yarnal and Leathers, 1988), using 1951 to 1960 and 1964 to 1973, respectively, to represent the two circulation regimes. Although he found differences between the two periods, the map-pattern frequencies do not do a good job in explaining interannual and interdecadal temperature variations. For instance, his map-pattern frequencies predict a 1.2 °C warming from the zonal to the meridional decade, when in reality the Arctic experienced a dramatic cooling. Barry *et al.* (1981b) also used the correlation-based technique to study interannual and interdecadal climatic variability over the western United States. Their results were weak because some map patterns show considerable within-group variability over time. They concluded that frequency is only one component of interannual climatic variability; investigators must also account for within-group variation. Thus, each of the above studies found that within-group variation is a critical problem in synoptic-climatological analysis of climatic variability.

In a study of Pacific Northwest coastal climatic variability, Yarnal (1985a) found that wintertime precipitation totals go up when seasonal temperature increases along the Alaskan panhandle. Simultaneously, when winter temperature rises on the Washington coast, precipitation totals fall. The opposite precipitation responses occur when wintertime temperatures are below normal in both areas. Yarnal and Diaz (1986) demonstrated that the global-scale Pacific–North American (PNA) teleconnection pattern forces these variations in the shortwave weather systems and regional climate. More important to synoptic-climatological theory and method was Yarnal's explanation for the problems with the Bradley and England (1979), Keen (1980), and Barry et al. (1981b) analyses described in the last paragraph. Yarnal demonstrated that severe within-group variability results from mixing data from two different circulation regimes. Within each regime, within-group variability is relatively small and map-pattern frequencies account for most of the interannual variation in surface climate. Therefore, long-term synoptic climatologies must identify the break points between regimes and avoid mixing the two populations.

Using a correlation-based classification, Sharon and Ronberg (1988) focused on the associations between global-scale circulation systems and the variations in the synoptic-scale winter storms that strike Israel. The authors demonstrated that the number of "Cyprus lows" coming over or near Israel varies with the position of the Mediterranean trough. When the trough line is to the west, these lows fail to reach Israel because they either die in mid-Mediterranean or pass well to the north. When the trough is over the eastern Mediterranean, the storm track lies over Israel or just slightly north.

Although investigators usually apply synoptic climatology to mid-latitude and high latitude environments, two studies used the correlation-based method with tropical precipitation. Singh et al. (1978) identified the major flow patterns occurring during the Indian southwest monsoon. The aim of the study was to determine whether synoptic climatology presents an improvement over persistence when forecasting precipitation probability. The results showed that persistence was superior to the synoptic climatology for forecasting precipitation probability for (1) the next day in all regions and (2) two to four days for certain regions and selected rainfall criteria. Nevertheless, for all regions of the subcontinent, synoptic climatology was better than persistence as an aid to forecasting precipitation probability for periods of four days or greater. In a study that falls outside the working definition of synoptic climatology used in this book, Nicholson (1979) applied the correlation-based technique to annual rainfall departures over north Africa in an attempt to find spatial and temporal patterns in the departure fields. She classified 54 of the 73 years into seven distinct rainfall-anomaly patterns. Her findings suggest that associations between rainfall in the Sahel and other parts of Africa are not consistent. Also, the results point out that there is no simple explanation of Sahelian drought (for example, weakened intensity of the Intertropical Convergence Zone) and that investigators must explore complex arguments on both global and synoptic scales.

Suckling and Hay (1978) carried out a correlation-based synoptic climatology of solar radiation regimes over southwestern Canada. Their objective was to use the relationships between the atmospheric circulation and radiation data at a few points to develop a region-wide solar-radiation catalog. They found that although the map patterns did discriminate the spatial and temporal characteristics of the solar-radiation data, this discrimination was not good enough to meet their goal. They concluded that the inadequacies of their results can probably be traced to the inappropriate use of the typing techniques and inadequate synoptic data.

Barry *et al.* (1981a) used the correlation-based technique to look back in time. They studied modern map patterns to determine airflow patterns favoring abnormally high influxes of exotic pollen into the Baffin Island area. This knowledge is valuable because Quaternary specialists use pollen records to date the glacial stratigraphy and reconstruct the Holocene climate of the eastern Canadian Arctic. Their analysis suggests that areas south and east of Hudson Bay are likely to be exotic-pollen source regions much more often than Labrador–Ungava. However, because of the much greater distances, the more southerly air-parcel trajectories will have significantly lower pollen sedimentation rates. For these patterns of transport to account for the observed Holocene pollen spikes, the frequency of map patterns with southerly airflows would need to increase by an order of magnitude; this is impossible given the eight-week period of pollen dispersal among the species used for analysis. These results suggest that other factors, such as variable pollen production rates or a non-linear relationship between map-pattern frequencies and pollen deposition, must be responsible for the observed spikes in the pollen record.

Snow-and-ice environments are the most studied with the correlation technique, mostly because of the interests of R.G. Barry and his students — Moritz, Bradley, Keen, and Crane — whose Arctic studies were mentioned above. In another snow-and-ice study done under Barry, Crane (1978) established the synoptic controls on variations in sea-ice advance and retreat in the Davis Strait–Labrador Sea. Using Keen's (1980) catalog, he determined map-pattern frequencies for the June-to-August ice-retreat period, and the mid-October-to-November ice-advance period. The results show that increased frequencies of northerly and westerly airflows associate with late retreat and early sea-ice advance, while more frequent southerly wind trajectories relate to early retreat and late advance.

There have been a few correlation-based projects on ice and snow not initiated by Barry. Fitzharris (1981) found that major and minor avalanche winters in Roger's Pass, British Columbia share higher-than-normal frequencies of certain map patterns. He suggested that subtle distinctions in the sequencing of these critical patterns probably account for the differences between major and minor avalanche winters, but he did not attempt to deal with this problem. Petzold (1982) wished to study the energy and mass fluxes associated with regional lake break-up and freeze-up in Labrador–Quebec. However, he tried to use correlation-based synoptic climatology to predict relevant surface quantities for the region because surface environmental data are sparse. Unfortunately, his map patterns fail to discriminate radiation and energy adequately to provide daily estimates of these fluxes, although the weather patterns do predict evaporative flux reasonably well. Thus, his synoptic climatology is of limited use for his stated purpose. Yarnal (1984a; 1984c; 1984d) studied the relationships between 500-mb atmospheric circulation patterns and glacier mass balance in southwestern

Canada. He determined that (1) small-scale, high-wavenumber synoptic systems control the mass balance of the coastal glaciers of British Columbia, and (2) the continental glaciers of the Canadian Rockies respond to variations in the middle-tropospheric planetary waves. It appears that the rugged coastal terrain first lifts and then breaks apart the small low-level storms along the Pacific coast, effectively filtering this end of the atmospheric spectrum.

Turning away from surface environments to the atmosphere, Barry *et al.* (1987) hypothesized that the large-scale advection and moisture convergence associated with synoptic systems must cause the rapid increase in cloudiness during the Arctic spring. They combined a satellite-derived cloud climatology with a correlation-based classification of surface pressure to address this hypothesis. They uncovered a strong relationship between the formation of middle clouds and synoptic processes. Because high and middle clouds obscure low clouds, the relationship of low clouds to synoptic systems is uncertain, although airflow patterns determine whether low-level clouds will form in particular regions. In other atmospheric research, Robinson and Boyle (1979) applied the correlation-based technique to daily carbon monoxide (CO) and total suspended particle (TSP) concentration data from stations in the St. Louis Regional Air Pollution Study (RAPS). All of the surface-pressure patterns associated with high TSP concentrations have low winds and high pressure, while low TSP values relate to high-wind, low-pressure map patterns. The CO analysis produced similar results, in that it is possible to associate different pressure patterns with distinct CO concentrations. It is interesting to note, however, that there is a different set of weather patterns associated with higher average concentrations of the two pollutants.

In summary, it is possible to use correlation-based map-pattern classification to analyze various environmental settings and questions. In Chapter 6, I apply the correlation-based technique to air pollution, crop yields, and hydrologic problems. In the next section, I develop that correlation-based classification.

Worked example

Procedure description

The correlation-based classification procedure followed the steps in Figure 3.1. First, the investigators selected the data for use with the classification program. For comparability with the weather maps used in Chapter 2's worked manual classification, and because the environmental scenarios were believed to be related to lower-tropospheric circulation, the investigators picked gridded surface-pressure data. These data come from the National Meteorological Center (NMC) 1,977-point Northern Hemisphere grids archived by the National Center for Atmospheric Research. Of the 3,652 days in the 10-year study period, 49 were missing, leaving 3,603 days for analysis.

After long discussion, the investigative team extracted a five-by-seven window from the NMC grid. The window covers the domain shown in Figure 3.3 and used throughout this book. This 35-point grid is not centered on western Pennsylvania. Instead, it is offset to the west and south because the majority of synoptic systems affecting the study area come from the southwest. A lesser, but important, quantity comes from the northwest. The number of grid points takes into consideration earlier findings. The investigators did not wish to filter short-wave features by using a larger spacing between grid points (Yarnal, 1984a), so they opted to accept the NMC grid spacing of approximately 350 km. At the same time, they wanted to cover a relatively large area, maximize the percentage of days classified, and minimize the number of map patterns. Larger areas produce greater numbers of grid points and, therefore, lower percentages of days classified and larger numbers of map patterns (Yarnal and White, 1987). Thus, with a fixed grid spacing, the ultimate decision was on the size of the domain; that is, the number of grid points. Review of the literature and experience suggested that the 35-point grid would serve the purposes of the research adequately. With 35 grid points and the domain's offset, the classification would capture storm systems coming from the lee of the Colorado Rockies, the Gulf of Mexico, or northern Great Plains.

For surface-pressure data, standardization of the grids is essential to remove the seasonal changes in absolute pressure and pressure-pattern intensity. Hence, the investigators applied formula 3.4 to the gridded data.

The team used an updated and modified version of Yarnal's procedure to perform the correlation-based classification (Yarnal, 1984b). Before starting, they determined the computer memory needed for classification. If memory requirements were too high, they would have had to devise a grid-sampling strategy. Using formula 3.5, they calculated that the 3,652 grids require nearly 6.7 M of RAM, not counting an additional few hundred K for processing the data. The RAM limit for normal users of the Penn State mainframe is 2 M, so the team obtained special permission for a higher limit (at a higher charge). By processing all of the data at once, the investigators avoided the problems associated with sampling in correlation-based classifications (Yarnal and White, 1987).

The investigators dealt with the challenge of setting thresholds the traditional way; that is, trial and error. After 13 test runs, they settled on $r = 0.4$ for the overall threshold. More stringent thresholds produced many map patterns with relatively small frequencies per pattern; smaller r-values produced just a few classes with lower-order map patterns that were too large (for example, the first map pattern might contain more than 50% of the days). Experimenting with the row and column thresholds, they found that decreasing the correlation values to 0.0 had no impact on the selection of keydays. Essentially, a sector threshold of zero means that this algorithm screens features of opposite sign but has no other effect.

The program calculated the grid-pairs and produced the keydays. Following the advice of Petzold (1982), the investigators set the minimum group size to one. With $r = 0.4$, this produced 43 keydays, including many with fewer than 37 members, 1% of the 10-year study period. It is important

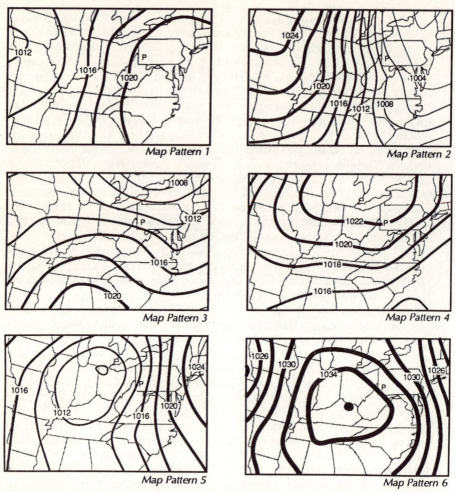

Figure 3.3 Correlation-based map-pattern classification: average sea-level pressure pattern (millibars) associated with the synoptic types.

to note that of the 12 keydays eventually retained, four of these had less than this 1% figure in the initial classification run.

Next the investigators reclassified the data using these 43 seeds. After this reclassification, the program only recognized 32 keydays. At this point the scree plot shown in Figure 3.2 was employed to choose the number of map patterns. The investigative team considered six likely breaks in slope. The first occurs after map pattern 4. This break is inappropriate because it includes just four map patterns and accounts for only 49% of the days. The second break stops at seven map patterns and uses almost 62% of the grids, figures the investigators again considered too small. Break three ups the numbers to nine map patterns and over 69% of the days, while the fourth change in slope includes 12 map patterns and roughly 78% of the days.

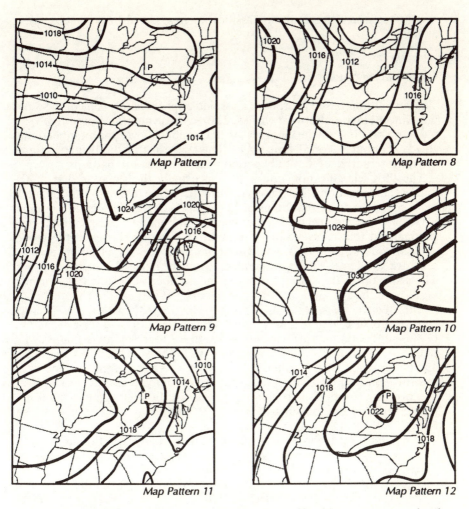

Figure 3.3 cont. Correlation-based map-pattern classification: average sea-level pressure pattern (millibars) associated with the synoptic types.

Breaks five and six increase the map-pattern total to 16 and 19, which the investigators considered too many. Thus, the choice was between nine and 12 map patterns. Studying the keydays, the investigators noted that the 10th, 11th and 12th keydays of the 12-pattern classification appear to be significant weather makers in western Pennsylvania. With this in mind, they selected the 12-pattern classification. Given the results of the worked example in Chapter 2 which was based on nine synoptic types, either choice would probably have produced satisfactory results. The uncertainty over the number of classes to use highlights the subjectivity of this and other synoptic-classification methods.

Based on these 12 keydays, the investigators reclassified the grids one last time. They stored the resulting classification in a spreadsheet for subsequent analysis.

Classification results

The correlation-based procedure classified approximately 95% of the 3,603 days into 12 map-pattern categories. There is little month-to-month or year-to-year variation in the number of unclassified days; the number holds steady at about 5% on all time scales. This suggests that the 12 map patterns are representative of the study period's surface-pressure patterns.

The 95% classification rate is almost exactly the same as that of the manual classification. This is not a coincidence. One reason the investigative team chose the scree-plot break at 12 (Figure 3.2) was because this number is close to the nine synoptic types used in the manual classification. They rationalized that using a similar number of synoptic classes in the model comparisons (Chapter 6) would determine if one classification is more effective at identifying representative synoptic categories than the other. The same rationale is used to choose the number of synotic types or map patterns in the synoptic-classification schemes of subsequent chapters.

The investigators decided to portray the 12 map patterns as average surface-pressure patterns, rather than keydays (Figure 3.3). Hoard and Lee (1986) suggested that mean patterns are more representative than keyday patterns, which are really the median days in the statistical distribution of any one map pattern. Furthermore, keyday maps represent just one day's pressure pattern and, therefore, can have individualistic quirks. Brief descriptions of the 12 map patterns follow, including a comparison to the manual classification scheme.

— Map Pattern 1 shows a strong Bermuda high over the study area. Assuming the normal westerly progression of weather systems throughout the day, the weak southerly flow will strengthen and veer to the southwest as a low-pressure center approaches from the west. This map pattern is associated with advection of moist tropical air and can be expected to bring elevated temperature and dew point to the study area. If the eastward progression is slower than normal or stalled, then these conditions can last for days. In the manual classification scheme, this is a back-of-high (BH) pattern.
— Map Pattern 2 depicts the recent passage of a low, with a steep pressure gradient and strong northwesterly wind blowing over the Great Lakes. This is the classic setup for lake-effect precipitation and would be categorized PH_w in the manual classification.
— The third map pattern shows troughs just to the east of the study area and over the midwest, connected by a low-pressure center north of the Great Lakes. Together with the intervening tongue of warm, moist tropical air extending from the southeast, this pattern suggests a classic cyclone passage, with its typical sequence of warm front, warm-air sector, and cold front. This is an RC pattern in the manual scheme.

— Map Pattern 4 shows a high centered over the Great Lakes. This PH_d look-alike typically generates lazy northerly flow, relatively low temperature, low dew-point temperature, and clear sky over the study area.

— A mature cyclone encroaches on the study area on Map Pattern 5 days. More than likely, an occluded front will subject the area to a lowering sky, steady precipitation and below-average temperature for a period of at least 24 hours. This is also an RC type in the manual classification.

— A massive high-pressure system is moving into the study area with Map Pattern 6. The high central pressure suggests that this is a winter pattern. (Although the correlation-base algorithm cannot identify pressure intensities, averaging the patterns together does facilitate such interpretations.) Thus, the high will produce moderate west-northwesterly flow with lake-effect cloud and precipitation. It would be classified PH_w.

— This is a difficult pattern to interpret because it does not fit the classic cyclone model. Map Pattern 7 has a meandering, possibly stationary frontal boundary extending across the Carolinas from the low over the Gulf states. A very weak, probably slow-moving front drapes across the Great Lakes from north of Lake Ontario into the midwest. Thus, although there is a high over the study area, it is surrounded by low pressure and weak fronts. Weather in western Pennsylvania is overcast and cool; light precipitation could occur at any time; and wind is light with no preferred direction. Upper-air charts would show a trough or cut-off low over the study area; therefore, this would be an extended low (EL) in the manual scheme.

— Map Pattern 8 implies a strong low-pressure center north of the Great Lakes with a northeasterly trajectory. An occluded front extends south from the low, connecting with the warm-air sector somewhere around the Ohio River Valley. The study area is probably experiencing veering easterly to southeasterly wind, cloudy sky and precipitation. Conditions will not improve considerably during the day. This is another RC type in the manual scheme.

— On Map Pattern 9 days, a strong coastal storm, known as a *Nor'easter* in the eastern United States, travels up the East Coast. Depending upon the low's position and strength, the study area may experience: (1) a "wrap around" effect from the low-pressure center, bringing northeasterly wind, cloud, cool air, and precipitation; (2) northwesterly flow off the Great Lakes with lake-effect precipitation; or (3) something in between, with northerly flow and little cloud and precipitation. In the manual classification, this would probably be classified PH_w, although closer passage of the storm or a stronger low with an expanded cloud shield could make it an RC.

— Map Pattern 10 has a low over or just north of, the Great Lakes, with a trailing cold front through the midwest. Over western Pennsylvania, conditions are those associated with a back-of-high pressure, but they will surely change to cold-front weather during the day. This is CF in the manual scheme, unless the next day's map shows that the CF slows or

stalls; then it is BH.
— The recent passage of a cold front ushers in a huge high-pressure dome on Map Pattern 11 days. The study area experiences pre-high (PH) conditions (northwesterly flow and perhaps lake-effect cloud and precipitation) in the early part of the day. However, if the high center covers western Pennsylvania later in the day, wind could slow, cloud could disappear, and temperature could rise dramatically from a combination of compression and radiation heating. With appropriate upper-level support, Map Pattern 11 highs sometimes stagnate: that is, they become EH types in the manual classification.
— Map Pattern 12 places a high over western Pennsylvania, with clear sky and relatively high temperature. Air flow is indeterminant. In the manual system, this is either BH, if the high migrates eastward quickly, or EH, if it stagnates over the area.

The most common map pattern in virtually every correlation-based classification is the first map pattern identified by the keyday algorithm. The results here are no exception, with the Map Pattern 1, Bermuda-high situation occurring about 22% of the time during the ten-year study period (Figure 3.4, top). The second most frequent category is a pre-high pressure pattern, Map Pattern 2, which appeared on nearly 18% of the days. The next two most common days are Map Patterns 3 and 4, which had frequencies of 11% and 7%, respectively. Map Patterns 5 through 12 ranged from 3.0% to 5.4%.

The major difference between the results of the correlation-based classification and the manual classification is the emphasis of the former on high pressure. The correlation-based scheme clearly favored high-pressure patterns, identifying them on over 60% of the days. In comparison, the manual classification classified at least 8%, and perhaps as many as 14% fewer high-pressure patterns. The reason for this discrepancy is not apparent, but it might be attributed to the investigative team's focus on features that generated precipitation over western Pennsylvania such as cyclonic systems.

The map patterns had distinct seasonal cycles. For instance, the identification of Map Pattern 1 as a Bermuda high is confirmed in Figure 3.4 (middle). The frequency of this feature nearly doubled during the May-to-November hurricane season, suggesting a northward shift of the circumpolar vortex. Similarly, the presumed association between Map Pattern 2 and lake-effect precipitation is implied by the fact that this pressure pattern occurs two to three times as often during the winter months. Map Pattern 3, a cyclonic pattern with the low-pressure center to the north, had a definite peak in the summertime, while cyclonic Map Pattern 5 and its more southerly low had depressed frequencies in the heart of summer, June through August. Again, this suggests a migration of the average position of the polar front. Map Pattern 6, with its high central pressure, showed an unambiguous spike in winter. Each of the other map patterns had physically reasonable forms of within-year variation (not shown).

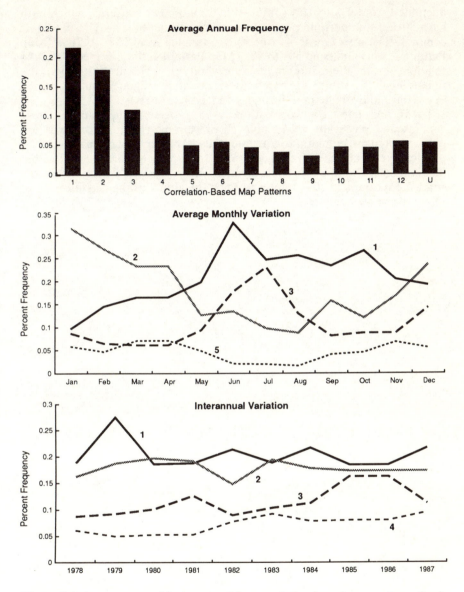

Figure 3.4 Average annual frequency of the correlation-based map-patterns (top), average monthly variation of four map patterns (middle), and interannual variation of the most frequent low- and high-pressure map patterns (bottom).

The map patterns also had considerable interannual variation. One of the least variable is usually Map Pattern 1, the Bermuda high, which hovers around 20% to 22% of the days (Figure 3.4, bottom). However, in 1979 this map pattern climbed to nearly 28%. In contrast, cyclonic Map Pattern 3 fluctuated from 9% to 16%. Interestingly, these two patterns, which both have strong summertime peaks, vary roughly in antiphase. The two most common lake-effect patterns varied less, ranging from 15% to 20% (Map Pattern 2; wet) and from 5% to 9% (Map Pattern 4, dry), respectively. No relationship in their interannual variability is evident. Although the remaining map patterns are much less frequent, averaging about 4% to 5% of the days, they fluctuated through a range of 2% to 8%.

To recapitulate, the correlation-based procedure appears to have generated a physically reasonable classification. The map patterns are plausible and are common to the area; they have seasonal cycles that make physical sense; and their frequencies vary from year to year, suggesting that fluctuations in map-pattern occurrences may produce variations in the surface environment. In Chapter 6, I test the ability of the classification to relate to the surface.

4 Eigenvector-based classifications

Classifications based on the eigenvector model were virtually impossible before the advent of computers. The thousands of hand calculations needed to perform an eigenvector analysis on a matrix of even moderate size prevented this technique from being utilized. Because investigators can now run such analyses on a mainframe computer in seconds or a desktop unit in minutes, eigenvector-based classifications have become common.

This chapter addresses three uses of eigenvector analysis to categorize data in synoptic climatology: synoptic-type classification, map-pattern classification, and regionalization. In the *Methodologies* section, I discuss the fundamentals of eigenvector analysis shared by all techniques and demonstrate the methodological differences among these three strategies. Following that, I review previous applications of these and other multivariate strategies in synoptic climatology in the *Previous studies* section. Finally, I present worked examples of eigenvector-based synoptic typing and map-pattern classification. One of the most important goals of these worked examples is to show how to make the output of these classifications easy to interpret. In so doing, I hope to "demystify" eigenvector-based classification and make it more useful for synoptic climatologists.

All three of the eigenvector-based schemes, synoptic typing, map-pattern classification, and regionalization, employ the circulation-to-environment approach to synoptic climatology (Figure 1.1). Because of this, their ability to relate to the surface environment can be evaluated by the model-performance statistics presented in Chapter 6.

Methodologies

Eigenvector analysis in synoptic climatology

There are many ways to conduct an eigenvector analysis. A bewildering number of subjective decisions confront investigators using eigenvector techniques, many of which will have dramatic impact on the results. In fact, the number of final solutions which satisfy the equations of the various eigenvector models is infinite (Richman, 1986). In this brief introduction to eigenvector analysis, I will emphasize the many decision-points facing the investigator and suggest appropriate choices in the context of synoptic climatology. I will not cover the mathematics of eigenvector procedures because they are readily available. For example, Preisendorfer (1988) presents a complete description of these techniques in the atmospheric sciences.

The steps in an eigenvector-based classification in synoptic climatology

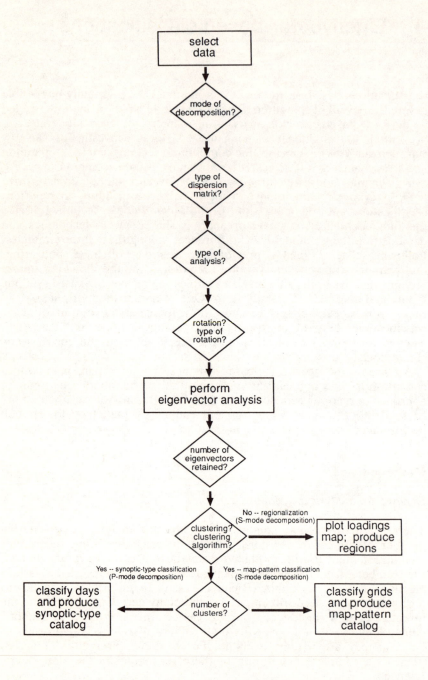

Figure 4.1 Eigenvector-based classification in synoptic climatology.

are flow-charted in Figure 4.1. In the first step, the investigator must choose the data for analysis (Figure 4.1: *SELECT DATA*). Ideally, the data should be continuous in time and space; missing data should be minimized. Investigators must not interpolate missing temporal data. They should execute spatial interpolation with care, especially over heterogeneous terrain. For analysis of a mapped surface, the data points should be evenly distributed in space. Investigators can interpolate mildly clustered data to a regular grid; they should never perform eigenvector analysis on severely clustered data.

An eigenvector analysis uses one of six modes of decomposition (Figure 4.1: *Modes of decomposition?*), defined as O, P, Q, R, S and T. In climatological applications of eigenvector analysis, there are three entities: (1) an atmospheric variable or field; (2) time; and (3) station (Richman, 1986). The investigator performs an analysis by varying two of these three entities and holding the third one fixed, producing one of the six modes. Figure 4.2 illustrates the possible configurations of the data matrix based on the arrangement of rows and columns, along with the dispersion matrix, the eigenvector-loadings matrix, and the eigenvector-scores matrix. In synoptic climatology, there are two common analysis types: P-mode, which analyzes a suite of variables varying over time; and S-mode, which concerns one variable (usually surface pressure or geopotential height) varying over space. Synoptic typing uses P-mode, while map-pattern classification and regionalization employ S-mode.

After selecting the data and preparing the matrix of observations, the investigator must choose the dispersion-matrix type for input to the analysis (Figure 4.1, *Type of dispersion matrix?*; Figure 4.2). There are three possible matrices: correlation, covariance or cross-products. Because a P-mode analysis uses more than one variable with more than one unit of measurement, the investigator must standardize the data and employ a correlation matrix. For S-mode analyses, the decision is more difficult. In S-mode, a correlation matrix produces standardized departure fields with non-dimensional isolines, just like the standardization process in a correlation-based analysis (Chapter 3). Thus, the comparisons reflect map-pattern shape but not intensity. A covariance matrix, on the other hand, more accurately depicts the actual spatial deviations of the data. However, a covariance matrix will concentrate pressure-field centers in the areas of maximum variance. This is a problem if there is a gradient in variance across the study area (Richman, 1981) which is often the case. Thus, based on these considerations of S-mode analysis, map-pattern classifications should use a correlation matrix, while a covariance matrix is better for climate regionalizations (White *et al.*, 1991). There has been little eigenvector-based climatological research using a cross-products matrix.

The investigator then must decide on which type of analysis to perform (Figure 4.1; *Type of analysis?*). Climatology employs three types: common factor analysis (CFA), principal components analysis (PCA) and empirical orthogonal function (EOF) analysis. All three come from the eigenvector model, in which the investigator interprets the underlying structure of a multidimensional data set by extracting eigenvalues and eigenvectors from

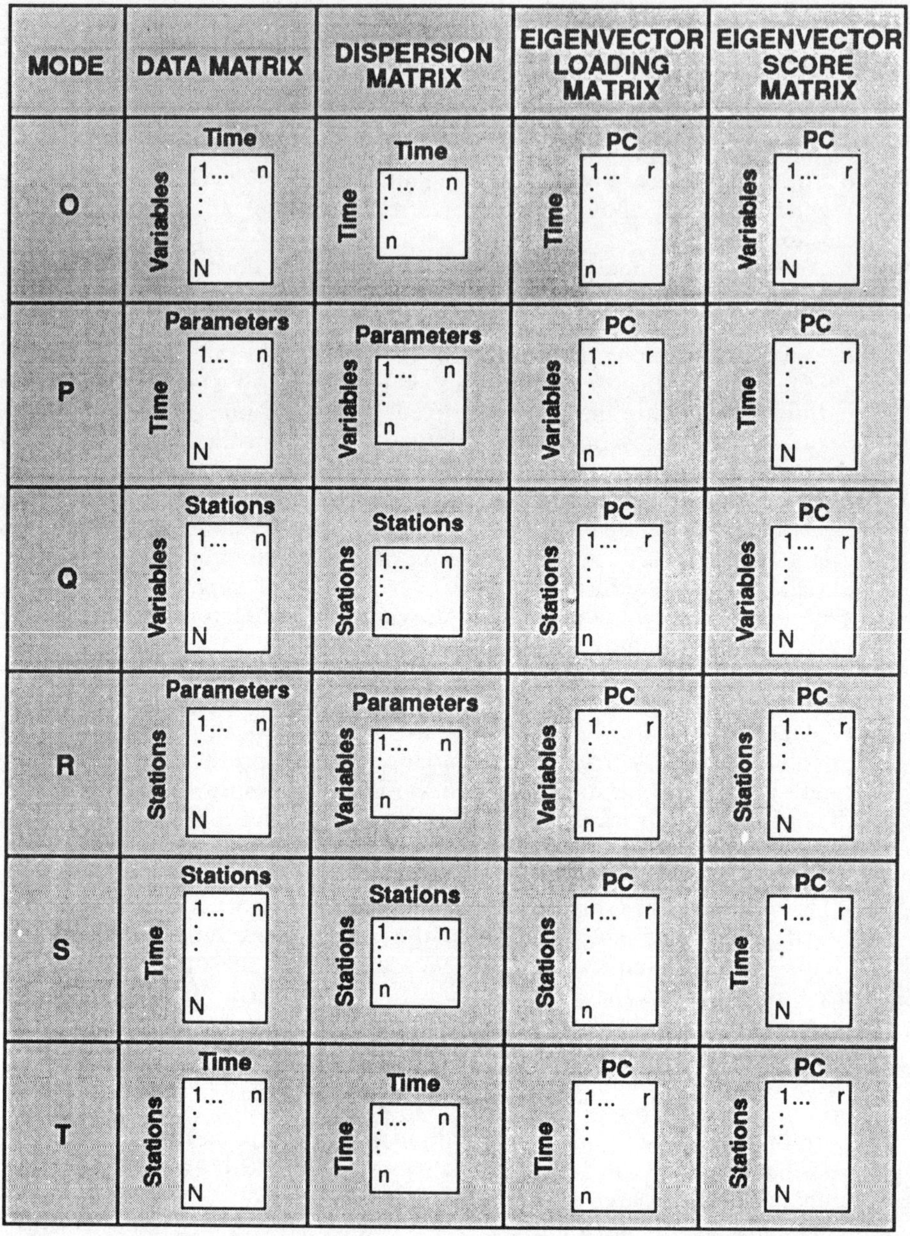

MODE	DATA MATRIX	DISPERSION MATRIX	EIGENVECTOR LOADING MATRIX	EIGENVECTOR SCORE MATRIX
O	Variables × Time: 1... n to N	Time × Time: 1... n to n	Time × PC: 1... r to n	Variables × PC: 1... r to N
P	Time × Parameters: 1... n to N	Variables × Parameters: 1... n to n	Variables × PC: 1... r to n	Time × PC: 1... r to N
Q	Variables × Stations: 1... n to N	Stations × Stations: 1... n to n	Stations × PC: 1... r to n	Variables × PC: 1... r to N
R	Stations × Parameters: 1... n to N	Variables × Parameters: 1... n to n	Variables × PC: 1... r to n	Stations × PC: 1... r to N
S	Time × Stations: 1... n to N	Stations × Stations: 1... n to n	Stations × PC: 1... r to n	Time × PC: 1... r to N
T	Stations × Time: 1... n to N	Time × Time: 1... n to n	Time × PC: 1... r to n	Stations × PC: 1... r to N

Figure 4.2 The six modes of decomposition in an eigenvector analysis and their data matrix, dispersion matrix, eigenvector-loadings matrix, and eigenvector-scores matrix (modified from Richman, 1986, with permission of the *International Journal of Climatology*, Royal Meteorological Society).

the dispersion matrix. Each eigenvector is orthogonal to (that is, uncorrelated with) every other eigenvector in multidimensional space. Thus, the eigenvectors isolate and define distinctive modes of variation in the data. Relaxing the orthogonality restraint means that more than one eigenvector will capture the data's variation. See Gould (1967) for an excellent intuitive, geometrical interpretation of eigenvalues and eigenvectors.

Methodologically, CFA differs from PCA and EOF analysis in one crucial step: determination of the level of communality. Communality refers to the amount of common variation explained by the eigenvectors; its compliment represents the proportion of the unique, unexplainable variance found in any variable. In CFA, each variable's unique variance is determined by one of many methods. To determine the level of communality when using a correlation matrix, the investigator subtracts the unique variance from unity; when employing a covariance matrix, he or she subtracts the unique variance from the total variance. Then the investigator inserts each variable's communality in the diagonal of the dispersion matrix. In contrast, PCA and EOF analysis assume that there is no unique variation in the data (the eigenvectors account for 100% of the variance); hence, the investigator does not change the diagonal elements of the dispersion matrix. Although this is an unrealistic way to view data, these models are far easier to apply and understand. An investigator should never use CFA unless he or she has a sound understanding of the processes operating on the observed data and the theory behind the CFA model.

Nevertheless, because CFA recognizes unique and common variance, it can be subjected to rigorous statistical testing. In contrast, the idealistic, common variance-only approach of PCA and EOF analysis cannot be addressed by probability and hypothesis testing; therefore, they are not true statistical techniques. Instead, they are mathematical manipulations that assume some of the characteristics of statistical procedures. Originally devised as methods of variable reduction, PCA and EOF analysis belong to that category of techniques "in which utility is judged by performance and not by theoretical considerations" (Davis, 1973, p. 501).

Since Lorenz (1956) introduced it to meteorology, EOF analysis has been the preferred eigenvector model of that discipline. An EOF is a unit-length eigenvector. In contrast, a principal component weights the eigenvectors by the square root of the corresponding eigenvalue so that these weights (known as loadings) represent the correlations (or covariances) between each variable and each principal component (Richman, 1986). Consequently, principal components convey more information than EOFs. The more elegant principal-components model is the favorite among climatologists working outside meteorology.

For classification, EOF analysis presents no advantage over PCA, while factor analysis is both difficult to use and to justify theoretically. Additionally, commercial PCA software is readily available for mainframe and desktop computers; this is not the case for EOF analysis. Therefore, in the remainder of this book, I will employ PCA and use its language (for example, I will refer to eigenvectors as components).

After deciding on the type of analysis to perform, the investigator must decide on whether a mathematical transformation of the orthogonal components, known as rotation, is appropriate or not (Figure 4.1: *Rotation? Type of rotation?*). Originally, investigators applied rotation to the eigenvector model because it facilitates interpretation of the principal components. More recently, they have discovered other benefits of rotation. There are two categories of rotation: orthogonal and oblique. Orthogonal rotations change the relationship between the various components but retain the orthogonality constraint of the eigenvector model. Oblique rotations relax this constraint, allowing some shared variance among the components.

Each option (that is, non-rotation, orthogonal rotation and oblique rotation) has its proponents. Investigators promulgating non-rotation point out that unrotated components extract the maximal variance from a data set. Richman (1986) suggests that investigators should apply unrotated solutions when seeking pure data reduction or when using the principal components for regression without interpretation. However, he believes that rotation is necessary in spatial contexts because "unrotated solutions exhibit four characteristics which hamper their utility to isolate individual modes of variation: . . . domain shape dependence, subdomain instability, sampling problems and inaccurate portrayal of the physical relationships embedded within the input matrix" (Richman, 1986, p. 295). Strict adherents of non-rotation object to rotation on the grounds that the unrotated analyses produce the most parsimonious solutions; rotation departs from these solutions and therefore degrades the results. Relaxing the orthogonality constraint in oblique rotations particularly appalls them. Supporters of rotation do not worry about mathematical elegance but instead are interested in providing the most interpretable outcomes. They contend that because PCA is not really a statistical procedure, theoretical considerations are secondary and utilitarian factors must dictate the use of rotation. Richman has been the chief proponent of principal-components rotation in the atmospheric sciences, demonstrating the vastly superior results produced by rotation in spatial contexts (for example, Richman, 1981 and 1986; White *et al.*, 1991).

To summarize the rotation problem in the context of synoptic climatology, synoptic typing, map-pattern classification, and regionalization require different strategies. In synoptic-type classification, PCA is primarily a data reduction technique and, therefore, the unrotated solution is appropriate; there is no advantage to rotating the principal components. In contrast, map-pattern classification and regionalization demand rotation because PCA identifies the spatial modes of variation over a mapped surface. However, as will be shown later, because it is necessary in a map-pattern classification to subject the outcome of the PCA to cluster analysis, the simpler orthogonal solution is preferred over oblique rotations. Oblique rotations provide better solutions than orthogonal rotations when regionalizing.

After collecting the data and deciding on the mode of decomposition, type of dispersion matrix, type of analysis, and whether to rotate or not and (if rotation is appropriate) type of rotation, the investigator can finally execute the PCA (Figure 4.1: *PERFORM EIGENVECTOR ANALYSIS*). There are three important products derived from the PCA: the principal components, the component-loadings matrix and the component-scores matrix. From the $n \times m$ dispersion matrix, the algorithm computes one principal component for each of the m original variables. The first component explains the most variation in the data, and each successive component describes a lesser amount of variation. An $m \times p$ component-loadings matrix is calculated and describes the way that each variable weighs (that is, loads) on each component (Figure 4.2). For any row in the component-loadings matrix, the sum of the squared loadings equals unity. Relationships between the variables and the components are not always clear but can be clarified by rotation. Indeed, rotation facilitates interpretation of the component-loadings matrix, making it easier to identify which variables load most highly on which components; unrotated solutions are often difficult to interpret. Multiplying the dispersion matrix by the loadings matrix produces the $n \times p$ component-scores matrix (Figure 4.2). Component scores illuminate relationships between the observations and the components.

PCA is primarily a variable-reduction technique. Although the algorithm produces m principal components explaining 100% of the data's variation, the first few components usually describe most of the variation. Hence, the investigator discards all but a few "important" components in a PCA (Figure 4.1: *Number of eigenvectors to retain?*). There are a number of approaches to determine the optimal number of components. Perhaps the most popular method is the scree test, in which a major break in a plot of eigennumber versus eigenvalue (the scree plot) indicates the appropriate quantity of components to retain. Another favored empirical technique eliminates all components with eigenvalues less than unity; this is intuitively appealing because the variation explained by each of the original variables is 1.0. A large number of more painstaking mathematical techniques are available (see Preisendorfer, 1988), although there is conflicting evidence as to whether these more accurately identify the "correct" number of components to retain than do the more empirical methods. White *et al.* (1991) suggest that the safest plan is, first, to employ several of these numerical and empirical techniques and, then, to make the decision on the number of components to retain based on the convergence of evidence.

This reduction in components truncates the component-loadings and component-scores matrices. However, the loadings and scores remain unchanged for the retained components. Both truncated matrices are important in synoptic-climatological classification.

At this point, the investigator must decide whether or not to apply a clustering procedure to the principal-components scores (Figure 4.1: *Clustering?*). If the object of the classification is regionalization, then the investigator does not cluster. Instead, he or she plots the principal-components loadings to produce regions. To recapitulate, when regionalizing, the investigator chooses S-mode decomposition, a covariance matrix, rotation, and no clustering.

If the investigator wishes to classify synoptic types or map patterns, then he or she subjects the component scores to a clustering procedure. Choosing the most appropriate clustering algorithm is critical to the outcome of the classification (Figure 4.1: *Clustering algorithm?*). For example, Kalkstein *et al*. (1987) compared three different clustering techniques (Ward's minimum variance, average linkage and centroid) in a synoptic-typing procedure. Each clustering algorithm produced vastly different synoptic types, with the average-linkage algorithm furnishing the most realistic results. Later in this chapter, I present a worked example of synoptic typing in which Ward's method provides better results than average linkage. As in the case of PCA and EOF analysis, the efficacy of any particular clustering procedure is judged by its performance.

The truncation point of the clustering procedure specifies the number of synoptic types or map patterns (Figure 4.1: *Number of clusters?*). Like PCA, there are many strategies for determining this cut-off, most of them empirical. Kalkstein and Corrigan (1986) used a scree plot of the coefficients of fusion, while Kalkstein *et al*. (1987) employed one for the squared multiple correlation (R^2), with breaks in the scree-plot slopes indicating points where unlike clusters were being forced together. The final number of clusters is arbitrary. It is usually based on an a priori, "ballpark" notion of the expected quantity of synoptic types or map patterns. Consequently, like the manual techniques described in Chapter 2 and the correlation-based method of Chapter 3, eigenvector-based classification provides as few or as many synoptic groupings as the investigator wants.

In this subsection, I discussed the rudiments of eigenvector analysis shared by synoptic-type, map-pattern, and regionalized classifications. In the following subsections, I address these three strategies to eigenvector-based synoptic classification more specifically.

Eigenvector-based synoptic-type classification

Synoptic typing attempts to capture the nature of and variations in the atmospheric circulation over time; spatial variations are not the primary concern. L.S. Kalkstein and collaborators at the University of Delaware have developed the synoptic-type strategy extensively.

After selecting the data, which might include such variables as temperature, dew-point temperature, wind direction and speed, cloud cover and visibility, the investigator must specify the mode of decomposition. Synoptic-type classification spotlights these variables' fluctuations over time at one point. Thus, P-mode is the appropriate configuration of the data matrix (Figures 4.1 and 4.2).

Next, the investigator must designate the type of dispersion matrix. Many of these variables use different scales of measurement (for example, degrees, meters per second, per cent, kilometers), thus requiring a correlation matrix.

Choosing the type of eigenvector analysis leaves a little more latitude than the previous two steps. Any one of the three analysis types, CFA, PCA or

EOF analysis, will produce satisfactory results. However, CFA is difficult to use and, without extraordinary a priori understanding of the variables and their interrelationships, is difficult to justify, making PCA and EOF analysis more appropriate choices. For continuity with this chapter's other sections, I will use PCA below and in the worked examples.

The investigator must now decide whether rotation is necessary or not. The large number of meteorological variables analyzed in a synoptic-typing scheme requires variable reduction. Thus, because PCA serves primarily as a variable-reduction technique in synoptic typing, there is no need to rotate the principal components.

The PCA produces the principal components, components-loadings matrix and components-scores matrix (Figure 4.2). One component will result for each of the original variables.

Once the investigator has run the analysis, he or she must determine the number of principal components to retain. The synoptic-typing classification technique appears to be quite robust, so that the determination of the number of components to retain is not as critical as in other climatological procedures such as regionalization (see White *et al.*, 1991). Because the first few components explain most of the variance, those of higher order are discarded, with truncated component-loadings and component-scores matrices remaining. The truncated component-loadings matrix describes the weather characteristics of each component, with the first one invariably loading highly on thermal attributes (Kalkstein, personal communication). The component scores depict the relative weights of each component on any given day. For example, the first day in the population may score highly on just component one (that is, a thermal component), while day two may score highly on both the second and third components (for example, humidity and visibility components).

The investigator then applies a clustering procedure to the daily component scores so that recurring assemblages of scores are recognized as representative. Both the specific characteristics and number of the synoptic types are dependent on two factors: the clustering algorithm applied to the scores and the truncation point of the clustering procedure. Kalkstein and collaborators (personal communication) have found that, with any one clustering algorithm, the size of the clusters varies with the climate type of the station. Thus, to produce realistic clusters, the investigator may have to change the algorithm with the environment. This requires prior knowledge of the synoptic climatology of the region and experimentation with various clustering procedures. Furthermore, the specific clustering procedure selected for the analysis and the cut-off point of the clustering algorithm will affect the number of synoptic types. No matter which clustering program the investigator uses, a scree plot shows when unlike clusters are being forced together and therefore determines the number of clusters (that is, synoptic types). Most scree plots will show several breaks in slope to choose from, and the investigator must make a subjective decision as to how many clusters are appropriate for the study period. It is critical to note that the point *before* the break in slope determines the number of types to retain. In the final appraisal, the results of a synoptic-type analysis are quite subjective.

The final step is to ascertain each day's cluster and to enter it into a synoptic-type catalog (Figure 4.1: *CLASSIFY DAYS AND PRODUCE SYNOPTIC-TYPE CATALOG*). The synoptic types are holistic ensembles of weather conditions at one station. They typically embody a particular air mass instead of necessarily being associated with any one map pattern. However, it is desirable to select a weather map to represent each synoptic type. This step is not necessary for utilizing the procedure, but it does help the investigator better understand the synoptic types. I discuss the interpretation of eigenvector-based synoptic types in the *Previous studies* and *Worked example* sections.

Eigenvector-based map-pattern classification

Map-pattern classification using eigenvectors produces spatial configurations that, at first glance, are similar to the results of the manual and correlation-based techniques discussed in earlier chapters. Eigenvector-loadings patterns, however, do not represent actual pressure surfaces; investigators must not interpret them as such. Instead, the loadings patterns for each eigenvector portray the main modes of variation in the pressure field. Furthermore, application of a clustering procedure to the pressure grids associates each grid with more than one eigenvector-loadings map. Thus, because the loading maps lack the interpretability of real weather maps and because the technique does not directly relate any given grid to a single keyday pattern, many investigators have shied away from this form of map-pattern classification.

The data selected for eigenvector-based map-pattern classification (Figure 4.1) are usually the same surface-pressure or pressure-surface grids used for correlation-based analysis. If these data are not gridded, the investigator should take care to insure that they are evenly dispersed over the mapped area. If a point-pattern analysis demonstrates no significant clustering in the data, the investigator can use them as is, or, for convenience, interpolate them to a grid. Significantly clustered data should not be utilized. Recovery of clustered point patterns is only possible if the investigator locates data for data-poor areas of the map.

After preparation of the data, the investigator selects the mode of decomposition, type of dispersion matrix, and type of analysis. Eigenvector-based map-pattern classification targets the main modes of spatial variation of just one variable; usually surface pressure or geopotential height. This is S-mode eigenvector analysis (Figure 4.2). S-mode map-pattern analysis uses the correlation matrix because variance gradients do not affect it, as in the case of the covariance matrix. Although the use of a correlation matrix eliminates the intensity of the pressure field, this is a benefit when using data that cross seasonal boundaries because the standardization procedure used to create the matrix removes the seasonal cycle. The investigator then enters the correlation matrix into the eigenvector analysis of choice which is usually a PCA.

The PCA produces component loadings and component-scores matrices, with the m principal components corresponding to the m data points on the map. Using empirical, mathematical, or a combination of these two methods, the investigator discards all but the most important components to determine the number of eigenvectors retained.

Buell (1975 and 1979) demonstrated that in an S-mode analysis, unrotated principal components manifest a regular sequence of loadings maps that is unrelated to the spatial variation in the data (see also, Richman, 1986, and White *et al.*, 1991). These patterns are simply statistical artifacts. Hence, it is necessary to rotate the components. The type of rotation should be orthogonal for two reasons: (1) the investigator clusters the daily component scores in the next stage of the classification, making the fine tuning provided by oblique rotation unnecessary; (2) the shared variance among the components in an oblique rotation prohibits the calculation of explained variance (R^2), which is a meaningful measure of each component's importance. Here it is customary to map the component loadings for each of the retained rotated principal components, with the maps portraying the spatial modes of variation for each principal component.

At this stage in the analysis, it would be ideal to assign the grids to a particular map-pattern class. This would be a simple task if each grid scored on just one principal component, or if the investigator chose the principal component on which a given grid scored highest as the map-pattern class for that day. However, each grid scores on all of the retained components, and the information carried in every individual component score is crucial. The variance field for each grid is an amalgamation of principal components, making the spatial pattern a complex expression of the scores' weights. Therefore, an almost infinite number of combinations of these weights is possible. The investigator applies a clustering algorithm to the scores matrix to identify the most common combinations of principal-component scores. To date, no one type of cluster analysis has emerged as the favorite for this purpose.

No matter which clustering algorithm the investigator selects, the determination of the number of clusters to retain is of considerable consequence. As in the case of the P-mode synoptic-type classifications, the investigator must study scree plots of various statistics generated by the cluster-analysis package. Significant breaks in slope on these plots indicate how many clusters best describe the data.

It is important to emphasize that there is no way to portray these clusters directly in map form (Crane and Barry, 1988). However, there is still an indirect way to picture the atmospheric configuration and weather associated with each cluster. Because each grid is assigned to a particular cluster, the investigator can average all the grids from each cluster to calculate average maps. These maps are physically interpretable and are just as useful as the output from any other automated synoptic climatology. I use this strategy in the worked example of eigenvector-based map-pattern classification presented later in this chapter.

Once the investigator determines the appropriate number of clusters and assigns each day to one of those categories, he or she produces a synoptic catalog (Figure 4.1: *CLASSIFY GRIDS AND PRODUCE MAP-PATTERN CATALOG*). This data array is usually entered into a spreadsheet for further analysis.

Eigenvector-based synoptic regionalization

Climatic regionalization using eigenvectors is not a technique exclusive to synoptic climatology. For instance, White *et al.* (1991) regionalized Pennsylvania's precipitation field using this approach. However, if this technique is first applied to a pressure surface to define anomaly patterns, and then related to the surface environment, it fits the working definition of synoptic climatology.

To perform an eigenvector-based synoptic regionalization, the investigator starts by selecting the atmospheric data (Figure 4.1). The easiest to use are gridded pressure-surface data such as those prepared for the eigenvector-based map-pattern classification. Regionalization demands S-mode decomposition (Figure 4.2), with the data matrix displaying n grid points by N observations of pressure.

Unlike synoptic typing and map-pattern classification, synoptic regionalization entails a covariance dispersion matrix. The covariance matrix accurately portrays the data's spatial deviations, concentrating the pressure centers in the areas of maximum variance (Richman, 1981). Thus, it picks out the anomaly fields.

Regionalization requires rotation of the principal components (Richman, 1986; White *et al.*, 1991). Although orthogonal rotation produces good results, superior regional definition is obtained using oblique rotation. This presents a disadvantage, however, if the investigation requires knowledge of explained variance; obliquely rotated components cannot produce this statistic. If knowing the explained variance is important, orthogonal rotation is a satisfactory fall-back solution.

After the investigator determines the m eigenvectors (that is, regions) to retain, the component loadings maps are plotted and the regions are produced. White *et al.* (1991) show that the 0.4 loading contour is usually a good region boundary. The investigator completes the synoptic climatology by relating these regions to surface environmental data at a point or to regionalized anomaly fields of environmental data.

Previous studies

This section reviews four types of eigenvector-based studies. First I cover synoptic-type classification. Next I consider map-pattern classifications. Third, I acquaint the reader with synoptic regionalizations. Finally, I present other multivariate angles on synoptic climatology.

Eigenvector-based synoptic-type classification

Synoptic typing is the most commonly applied eigenvector-based procedure in synoptic climatology. Kalkstein, his students, and other collaborators are the principal proponents of eigenvector-based synoptic typing. Their work has advanced in two parallel streams, methods and applications. Kalkstein and Corrigan (1986) covered most of the steps presented in Figure 4.1. They assembled 28 daily weather variables (seven variables recorded four times per day) over five winters (451 days) and created a 28-by-451 P-mode data matrix. They then converted this to a correlation matrix and subjected that to an unrotated PCA. Keeping only those components with an eigenvalue greater than 1.0, the investigators retained five principal components which explain 78% of the variance in the data. Ward's clustering technique applied to the 451-by-5 components-scores matrix yielded 10 synoptic types. To determine the air-mass characteristics of the synoptic types, they calculated means for each cluster's 28 weather variables. Finally, Kalkstein and Corrigan studied representative weather maps for each weather type to aid their understanding and interpretation of the results, especially in relation to SO_2 pollution. Kalkstein *et al.* (1987) used a similar version of the procedure and focused specifically on the problem of selecting the best clustering algorithm for the methodology. They determined that the average-linkage method is superior to the centroid and Ward's techniques, at least with the test data set from New Orleans, Louisiana. Davis and Kalkstein (1990a) developed a spatial version of the procedure. However, because its present form is limited by the huge data matrices P-mode analysis demands, this promising strategy is still in the developmental stage.

These continuing refinements to the classification procedure have enabled Kalkstein and affiliates to apply the procedure to an increasing variety of environmental problems. Kalkstein and Corrigan (1986) used the technique with wintertime SO_2 concentrations in Wilmington, Delaware. They found that "the most offensive synoptic type," an anticyclonic continental-polar air mass, encourages poor ventilation, strong subsidence, and the likelihood of inversion development. In fact, the results confirmed that all extended severe pollution episodes in Wilmington are associated with anticyclonic circulation. Applying the spatial modification to one year's data from the contiguous United States, Davis and Kalkstein (1990b) uncovered similar relationships for SO_2 and NO_2 at nine cities.

In 1989, the United States National Park Service and Environmental Protection Agency attributed the frequent wintertime haze at the Grand Canyon to a large coal-fired power plant, the Navajo Generating Station. Kalkstein and Webber (1990) used synoptic typing to evaluate relationships between the atmospheric circulation and air pollution at six sites in Utah, Nevada, and Arizona. They reasoned that if similar synoptic situations yielded low visibility simultaneously at the six sites, it is unlikely that a point source is responsible. Their exhaustive study showed that regional, rather than local-source, contributions are responsible for low visibilities at the Grand Canyon. They found that there was virtually no evidence in any of their analyses that suggested the Navajo Generating Station has a significant impact on Grand Canyon pollution levels. In a related study, Kalkstein *et al.* (1990b) concluded that the long-range transport of pollutants from southern California, populated areas of Arizona, and polluted regions of Mexico are the likely sources of air pollution in the Grand Canyon and at Bryce Canyon National Park.

Synoptic-typing formed the basis of a stochastic precipitation model for Philadelphia, Pennsylvania (McCabe *et al.*, 1989). The investigators found that the synoptic types identified weather conditions associated with varying frequencies, intensities, and amounts of precipitation. They then used the type frequencies to simulate precipitation stochastically. In all 100 simulations, estimated precipitation matched observations. From this, the investigators applied the synoptic-type frequencies to climate-change scenarios to simulate future precipitation.

Kalkstein *et al.* (1990a) turned their attention to the problem of greenhouse gas-induced global warming. General circulation models suggest that the strongest warming should occur in polar regions, especially in winter. Kalkstein *et al.* hypothesized that if warming is underway, then polar air masses should also be warming. Using their eigenvector-based synoptic-typing procedure with data from four North American Arctic locations, they discovered that the frequency of the coldest air masses decreased over the last several decades, while the frequencies of the warmest ones increased. Furthermore, the coldest air masses (those that form in the Arctic and suffer no modification from transport) warmed from 1 °C to 4 °C over this period. Unfortunately, the technique could not uncover evidence linking this air-mass warming to anthropogenic causes.

The threat of global warming also spurred Kalkstein (1992) to investigate the impacts of climatic variability on human mortality. He established that one synoptic type is associated with abnormally high mean daily mortality in summertime St. Louis, Missouri. Eight of the 10 top mortality days occurred when this hot, oppressive type was present. He showed that if this air mass stagnates over St. Louis, human mortality rises steadily throughout the episode. Because this potentially hazardous weather type possesses average to below-average air-pollution levels, Kalkstein concluded that day-to-day mortality is more sensitive to weather fluctuations than to air pollution.

Brazel *et al.* (1991) applied Kalkstein's procedure to energy balance data measured over an Alaskan glacier during a single ablation season. They found strong relationships between the energy balances and synoptic types derived from a remote station. This suggests that reconstruction of the glacier's long-term mass balance might be possible from meteorological data alone.

Other investigators have implemented the eigenvector-based synoptic-typing technique independent of the University of Delaware school. At about the same time that Kalkstein was beginning to develop his procedure (Kalkstein, 1979), Crane (1979) studied the synoptic controls on the energy-budget regime over ablating fast ice. He applied a fairly standard synoptic typing scheme (see Figure 4.1) to six energy-budget variables: conductive flux, Bowen ratio, Richardson number, sensible heat, net radiation, and latent heat. He identified four synoptic types, two for the early melt season and two for later in the melt season. In each part of these seasons, one type is associated with suppressed melt, while the other relates to enhanced melt. Crane showed representative surface weather maps for each synoptic type to support and enhance his interpretations of the clusters.

Maheras (1984) developed an eigenvector-based synoptic-type classification to study wintertime climatic variation in Thessaloniki, Greece. He defined winter as October to April and used a 10-year, 2,122-day data set to produce 11 weather types. Computer limitations forced him to classify each year individually and pool the results. From this, Maheras could describe the typical winter seasonal cycle and anomalous winters in terms of their frequencies and timing. He did not use representative weather maps to illustrate his findings.

Ezcurra *et al.* (1988) used synoptic typing to uncover the relationships between the atmospheric circulation and acid rain in Spanish Basque country. They first measured nine chemical attributes of each rain event during one year. They subjected these data to PCA to determine the sources of variation (that is, pollution) in the rainfall samples. The investigators then ran the output of the PCA through a clustering algorithm to produce four "synoptic types" of their rainfall chemistry. Finally, subjective analysis of 850 maps suggested different air masses and airflow trajectories for each cluster. Those are:

— Continental, northeasterly;
— Marine, northwesterly;
— "Iberian," southwesterly;
— "Local," stagnant air.

These results suggest that the worst acid rain comes from the industrialized areas of France and the Basque country northeast of the station, moderately acidic precipitation is associated with air from Portugal, Spain and local sources, and the least acid rainfall from the Bay of Biscay.

In a slight variation of the steps shown in Figure 4.1, Stone (1989) classified twice-daily weather data for 15 years at Brisbane, Australia. He claimed to be using 27 variables, but four of those were the year, month, day, and hour, while eight were "occurrence or non-occurrence" of the cardinal wind directions. Deviating from the standard procedure, he tried one oblique and two orthogonal rotations to the PCA, settling on the Varimax orthogonal rotation as the most interpretable. He overfactored (that is, retained too many components) which tends to produce unacceptable results in an orthogonally rotated PCA (Richman, 1981). Finally, he clustered the results of the original cluster analysis, a practice gaining acceptance among synoptic-typing advocates (for example, Kalkstein *et al.*, 1990). Using a combination of inspection of surface synoptic charts and break-points on the cluster tree, he chose nine broad weather types, which he further discriminated to produce 25 mid-level subtypes and 63 high-level subtypes. The 63 subtypes usually associate with a particular map pattern. Stone did not relate the synoptic types to the surface environment or to climatic variability, although this is the ultimate aim of his research.

Todhunter (1989) applied Kalkstein's procedure to an urban energy-balance problem. He determined six synoptic weather types for the Boston area to drive an urban energy-budget simulation model. He used full-year data sets but, unfortunately, he neglected to remove the seasonal cycle from the data. Instead of identifying the daily air-mass changes, the resulting synoptic types essentially reproduce the generalized annual cycle of air masses: that is, summer continental, summer marine, winter continental, winter marine, mixed (liquid precipitation), and mixed (frozen precipitation). Thus, the resulting classification only identifies whether the surface station is in continental or maritime air mass, or under the front separating the two. It does not supply sufficient discrimination of daily weather conditions for most purposes. In the worked example of eigenvector-based synoptic typing, I discuss the critical problem of seasonal-cycle removal and demonstrate a solution.

Eigenvector-based map-pattern classification

Map-pattern classifications based on eigenvectors constitute a tiny group. A few investigators, such as Richman (1981) and Key and Crane (1986), have addressed eigenvector-based map-pattern methodologies. I used many of their ideas to develop the procedural model presented in the *Methodologies* section. Unfortunately, Richman (1981) did not clearly differentiate map-pattern classification from climate regionalization, severely limiting the utility of his findings. He did, however, unequivocally separate these two S-mode analysis types in Richman (1986). Key and Crane focused on the subjectivity in the correlation-based and eigenvector-based models. They determined that decisions by the investigator, such as the number of components retained or type of clustering algorithm employed, affect the results. They concluded that investigators must be thoroughly familiar with the classification technique, as well as the data and climatology of a region, to derive the best possible eigenvector solution. Richman (1981) and Key and Crane did not deal with climatological applications of the methodology.

Blasing (1975 and 1979) also compared correlation-based and eigenvector-based map-pattern classification schemes (see Chapter 3). He concluded that correlation techniques are superior to eigenvector methods for the following reasons: they are computationally easier and faster; they are free of orthogonality and phase constraints; and the investigator can assess the statistical significance of each data point's anomalies when using the correlation method, but not with spatial eigenvector strategies. Hence, he used his version of the correlation-based technique for synoptic climatological applications.

Lamentably, there are no examples in the literature in which an eigenvector-based map-pattern classification is related to surface climate. To fill this void, I develop a map-pattern classification using this procedure in the worked example later in this chapter. In Chapter 6, I demonstrate that this eigenvector-based classification produces results comparable to manual and correlation-based techniques.

Eigenvector-based synoptic regionalization

A handful of investigators have performed an S-mode analysis on the covariance matrix (that is, a regionalization) of pressure data and then compared the resulting regions to surface climate. They identified these regions as anomaly fields. Kiess and Riordan (1987) applied unrotated PCA to gridded surface-pressure data. Their resulting map patterns followed the classical Buell sequence associated with unrotated S-mode analysis. Thus, all conclusions based on the bogus anomaly fields were spurious.

In a more sophisticated, methodologically sound analysis, Whetton (1988) related monthly pressure-pattern anomaly fields to monthly surface-rainfall patterns over southeastern Australia. He tried to regionalize unrotated components, but found Buell patterns. He chose orthogonal rotation over oblique rotation because he believed the forced independence of the orthogonally rotated components would produce better correlations between the rainfall and pressure anomalies than the not-necessarily-independent oblique components. His results suggest that rainfall variations over southeastern Australia are strongly determined by topography and that each rainfall region's variability is a unique response to the large-scale atmospheric circulation.

In another synoptic climatology relating the results of two separate eigenvector-based regionalizations, Michaels (1985) compared anomalous mid-tropospheric heights with thunderstorm patterns over Florida. Unfortunately, the investigator did not report how he performed the PCA. Which dispersion matrix was used: was this a map-pattern analysis or regionalization? Were the components subjected to rotation or not? If so, did the investigator use orthogonal or oblique rotations? From clues, the reader can guess the form of the analysis. Michaels probably used a covariance matrix because anomalous height fields indicate regionalizations rather than map-pattern classifications. Unrotated components are likely. The first component of each classification explains a huge amount of the variance (for example, 63% in the mid-tropospheric anomaly field), suggesting an unrotated analysis. Furthermore, the components for the mid-tropospheric anomaly field display the classic Buell sequence, again implying non-rotation. It is more difficult to discern Buell patterns in the thunderstorm components because of Florida's odd shape. In summary, it is hard to judge whether this analysis is valid and if the results are reliable with the information provided. *Investigators must provide basic information on the mode of decomposition, dispersion matrix, rotation type, and any other subjective decisions they make in an eigenvector analysis.* Even knowledgeable readers cannot assess the results without this information.

Other multivariate strategies

The infinity of ways in which the investigator can implement the eigenvector model have produced an assortment of eigenvector-based synoptic climatologies. Other multivariate statistical techniques not based on the eigenvector model, such as cluster or discriminant analysis, have also been used to solve synoptic climatological problems. In this section, I review some of these diverse strategies.

In an early eigenvector-based synoptic climatology, Christensen and Bryson (1966) performed a P-mode synoptic-type classification. The investigators' plan differs from the synoptic-typing procedure in the following ways: they used an orthogonal rotation; they regressed each day's retained components on the original observations; and they classified each day's regression coefficients with a variant of Lund's (1963) correlation methodology. Although they interpreted the results in terms of the classic cyclone model, their results are roughly comparable to those of Kalkstein's more refined procedure. Ladd and Driscol (1980) applied the Christensen and Bryson classification methodology to radar echoes in Texas's high plains. They tried to reproduce the methodology faithfully, except that they used upper air, as well as surface observations. They also developed a manual classification to check the results of the automated procedure. Ladd and Driscol claimed the resulting eigenvector-based synoptic types are climatologically valid and superior to the manual classification, but they did not test their computer-based output to see if the categories are statistically significant. The classification discriminates the radar echoes reasonably well.

S-mode analyses produce regionalizations (when used with a covariance matrix) and map-pattern classifications (when used with a correlation matrix). For best results, the former should employ oblique rotation, while the latter should use orthogonal rotation. In map-pattern classification, the investigator clusters the grids to account for the contribution of each principal component to each grid point's variance. Crane and Barry (1988) used S-mode analysis with the unusual combination of a correlation matrix and oblique rotation; they did not cluster the component scores for each of the individual grids. This is an appropriate use of the eigenvector model if the goal of the investigation is to define the main modes of spatial variation and not to classify each grid. However, because each day's grid cannot be classified, the atmospheric circulation cannot be related to surface climate and, therefore, this does not meet the working definition of synoptic climatology used in this book.

It is possible to compare the joint spatial and temporal variation of the atmospheric circulation and surface environment without performing separate eigenvector analyses. Keables (1988) sought to identify spatial associations of North American mid-tropospheric circulation anomalies, upper Mississippi River basin precipitation, and discharge from the basin. He constructed a single data matrix containing warm-season data and representing each of these environmental elements distributed over time and space. An unrotated PCA identified the dominant modes of joint variability in these data. For the various circulation anomaly patterns, the investigator then calculated the occurrence probabilities of different precipitation-frequency and maximum stream-discharge classes. Twelve components were significant, accounting for nearly 73% of the total circulation-hydrology variance. Because of the non-rotation, the first component accounts for about 20% of the variance, and subsequent components account for an average of roughly 5%. Keables concluded that heavy precipitation and high discharge are most probable during spring and summer months in which there is anomalous southwesterly or southerly air flow over the basin. Below-average rainfall and discharges are highly probable during months in which the anomalous circulation over the basin is from the northeast or the north. The probabilities of high rainfall and high discharge are low when high pressure sits over the basin.

McCutchan and Schroeder (1973) and McCutchan (1978 and 1980) used stepwise discriminant analysis to classify synoptic weather types associated with southern California forest fires and surface-ozone pollution. In the earlier two studies, McCutchan manually classified the climate into five categories; in the third, he used cluster analysis to determine these five classes. He first constructed the technique to identify the five climate classes, then he used discriminant analysis to categorize independent data into one of these classes. In each study, significant differences existed in the fire or oxidant-exposure values among the classes. Similarly, Diab *et al.* (1991) applied discriminant analysis to manually determined synoptic types to see if they are statistically distinct. They concluded that four major rainfall-producing systems and four minor systems contribute significantly to the integrated annual rainfall over Natal.

Cluster analysis is rarely used by itself for synoptic-climatological analysis. Although employed often, it is almost always an adjunct to other methods. However, a few studies that rely solely on cluster analysis come to light. Kruizinga (1979) and Tsui and Lam (1979) presented early attempts to classify synoptic types and map patterns, respectively, using this multivariate strategy. Ambrozy *et al.* (1984) constructed cluster-based map patterns comparable in scale to the *Grosswetterlagen*, to which they compared their results. Sanchez *et al.* (1990) related meteorological clusters to air pollution over Spain. They employed 500-mb charts associated with the clusters to aid the interpretations of their clusters. Perhaps synoptic climatologists do not use cluster analysis by itself because they are a visually oriented group of scientists; they appreciate the holistic aspects of cluster-based classifications but need a weather map or some other visual device to grasp the nature of a synoptic-type category.

A worked example of eigenvector-based synoptic typing

Procedure description

The investigative team followed the sequence of steps for eigenvector-based synoptic typing depicted in Figure 4.1. They selected NOAA/National Climatic Data Center TD-1440 surface observations at the Pittsburgh International Airport for the analysis. These data consisted of seven variables: temperature, dew-point temperature, sea-level pressure, wind speed, wind direction, percentage cloud cover, and visibility. Observers collect these data four times daily at the synoptic hours. Thus, seven variables measured four times per day means that the investigators had 28 values for each of the 10-year study period's 3,652 days.

Because this was a full-year classification, seasonal cycles had to be removed from the data. Without deseasoning, the resulting synoptic types would only identify gross seasonal characteristics, for example, summer mT (maritime tropical) or winter cP (continental polar) air masses (Todhunter, 1989). A spectral analysis suggested that the maximum duration of synoptic-scale systems over the area is 13 days. Thus, applying a 13-day running-mean filter to the data removes variance longer than the synoptic time scale such as the seasonal signal. The filtered data represent a daily deviation around a moving 13-day average. For the small number of missing observations, investigators substituted means of the 13-day window centered on the missing day. See Hewitson and Crane (1992a) for further details of the spectral analysis and filter.

In an eigenvector-based synoptic typing, P-mode decomposition is appropriate. In this example, the 28 deseasoned variables (parameters) by the 3,652 days (time) formed the data matrix. Based on instructions from the investigators, the computer program transformed the data matrix into a correlation matrix and entered it into an unrotated PCA.

The PCA produced 28 components, one for each variable. Eight of these had eigenvalues greater than one, a threshold commonly used to determine the number of components to retain. Still, experimentation showed that the more components retained, the more the clustering algorithm struggled; that is, much time and money was needed to produce the synoptic-type clusters. Also, results did not improve noticeably with the retention of more components. Therefore, based on these facts and a scree plot of the component eigenvalues, five unrotated components, explaining a minimum of 5% of the variance each and a total of 62% of the variance, were retained.

Eigenvector-based synoptic typing requires clustering of the principal-components scores from the retained components. Kalkstein *et al.* (1987) showed that the average-linkage clustering algorithm did best in their treatment of TD-1440 data from New Orleans. Yet, in this case, the average-linkage algorithm produced results inferior to Ward's method. Perhaps the use of year-round data or the application of the 13-day filter either degraded the performance of the average-linkage clustering or improved the results of Ward's method.

The number of clusters to retain was not immediately clear. Commercial statistics programs provide many measures of cluster performance; the investigators used the SAS package, which offers several (SAS Institute, Inc., 1985). Each produced similar but often subtly differing results. The R^2 measure, which compares the sum of the squared variation within clusters to the total squared variation, is one of those metrics. This ratio varies from one (when the number of clusters equals the number of individuals; in this instance the 3,652 groups of five components) to zero (when the data form one giant cluster). The values can be scree plotted to reveal break points in the clustering process. The team chose R^2 because it is easy to understand and because Kalkstein *et al.* (1987) used it. A scree plot suggested that several different break points were possible, from just a few clusters to more than 20. The team settled on 11 clusters (synoptic types) for three reasons: (1) it was a major break point; (2) the results of the preceding manual classification and correlation-based classification suggested 11 was an appropriate number of types; and (3) results of an 11-type classification would be directly comparable to the manual and correlation-based results.

Classification results

The 11 eigenvector-based synoptic-type clusters form discrete climatic situations. Each type's typical weather (Figure 4.3) and average pressure pattern (Figure 4.4) display a unique synoptic climatology. The top panel of Figure 4.5 shows the percentage frequencies of the synoptic types, while the middle panel shows the annual cycle of the four most common types. In the following, I present a description of the synoptic types based on these figures.

Type 1 is the most common pattern, occurring nearly 23% of the time. It is characterized by a broad, diffuse high-pressure field with light westerly wind in Pittsburgh. Pressure rises throughout the daylight hours as the center of the high moves over the city; pressure falls at the end of the day as the center of the high moves east. The temperature shown in Figure 4.3 is relatively high when compared to the other types, suggesting that Type 1 is primarily a warm-season feature. The monthly frequency curve illustrated in the middle panel of Figure 4.5 corroborates this fact. Temperature sinks in the morning hours with long-wave radiation losses but jumps with solar heating and compressional warming throughout the day. Dew-point temperature and the moderate cloud cover drop as the high migrates over the region. Visibility, which is moderate to good overall, is lowest near sunrise but increases as relative humidity falls during the day. This is a fair-weather type.

The second type identified by the procedure is in many ways a continuation of Type 1 as it translates over the study area. While the high moves offshore, the following predictable events occur on Type-2 days: pressure falls; the wind swings from east to southwest; the advection of maritime-tropical air causes dew-point temperature to climb; in the morning, temperature rises with the advection of warm Gulf air but drops later in the day with the increase of relative humidity and cloud cover; and visibility decreases as humidity and cloudiness rise. Again, Type 2 is most important in the warmer months.

The station model of Type 3, another category with a distinct summer maximum, appears to portray the early morning passage of a warm front, occluded front, or low-pressure center. This continues the sequence started with Type 1. Pressure rises and levels at the end of the day, suggesting the approach of the center of the high as low pressure moves eastward. Wind is moderate and consistently west-southwest. After an initial increase, dew point sinks throughout the day, while air temperature consistently decreases. Note that the average temperature is highest with this type. Cloud cover is high, but clearing continues throughout the day. Visibility is

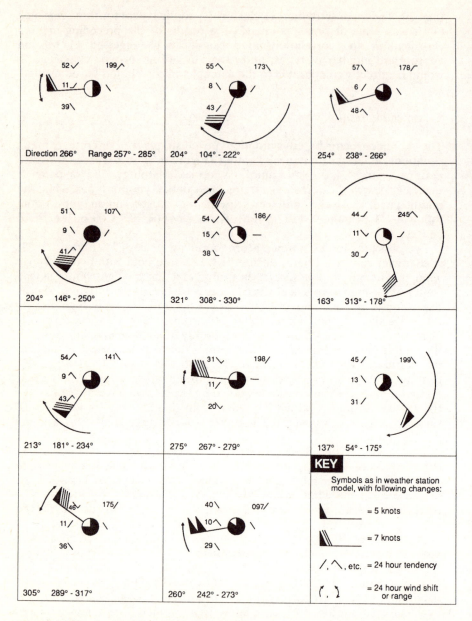

Figure 4.3 Station models representing the average weather conditions experienced with each of the eigenvector-based synoptic types.

lowest of all the synoptic types but improves continuously. The only inconsistency in this picture is the average pressure pattern, which, with normal westerly progress of the systems, implies falling pressure. The diffuse pressure pattern suggests that either pressure gradients are not steep with this synoptic type, or that Type 3 mixes several different pressure-pattern types, or both.

Type 5 depicts the typical pre-high conditions following a frontal passage. Pressure rises as the low exits and the high approaches. The wind is strong northwesterly. Dew-point temperature decreases as the heart of the dryer air mass arrives. Temperature falls in early morning with radiation losses but goes up with strong solar receipts. The sky is clear and visibility is the best of all the synoptic types. Temperature is high for a northwesterly type, implying more frequent occurrence of this type in summer.

To summarize to this point, Types 1, 2, 3 and 5 are the only categories that have their peak frequencies in summer. Except for Type 5, these types are also somewhat common during the cooler months, and the four types together account for over 60% of the study period's days. The large numbers and weak pressure gradients of summer explain why these four types have the most diffuse patterns in Figure 4.4. The remaining types, discussed below, are primarily cool-season phenomena with sharp pressure gradients and well-defined pressure patterns. Types 1, 2, 3 and 5 roughly depict the sequence associated with the passage of a wave cyclone.

Type 4 has its peak frequencies in the period November through March. Pressure falls throughout the day as a pronounced low approaches from the Great Lakes. In fact, pressure is dramatically lower on Type-4 days than on all but one other type day. Wind strengthens and shifts from southeast to west-southwest. Dew point climbs throughout the day, only to slide at its end, suggesting passage of a front and a change of air masses. Cloud cover is almost complete but manages to increase as the front approaches. Temperature tumbles as clouds thicken and, finally, the cooler air mass enters the area. Visibility drops while relative humidity and cloud cover rise. Note that temperatures are relatively high with this type, despite its wintertime peak in frequency. This suggests that strong advection of warm, humid maritime-tropical air is important to this type throughout the year.

Generally, Type 6 is most frequent in the cooler portions of the year, but its peak frequencies occur in the autumn and spring. This type displays the highest pressures and lowest wind speeds of any type. The wind is light and variable, but does show a swing from northwest clockwise through southeast, on average, as the high moves through the area. Pressure rises as the center of the high approaches and starts to fall as it translates eastward. After sunrise, temperature climbs from solar heating and incipient advection of southerly air. Likewise, dew-point temperature goes up with southerly advection. Cloud cover is nominal but does increase modestly throughout the day. Visibility is moderate but decreases during the day under the stagnant conditions; a slight jump in visibility is experienced as winds pick up near day's end.

A classic back-of-high situation occurs on Type-7 days. As the cold front nears, pressure dives, the wind strengthens, temperature climbs, dew-point temperature rises, visibility lowers after sunrise, and cloud cover thickens. Finally, when the cold front passes at the end of the day, dew point and air temperature start to tumble. Although Type 7 is four to five times more

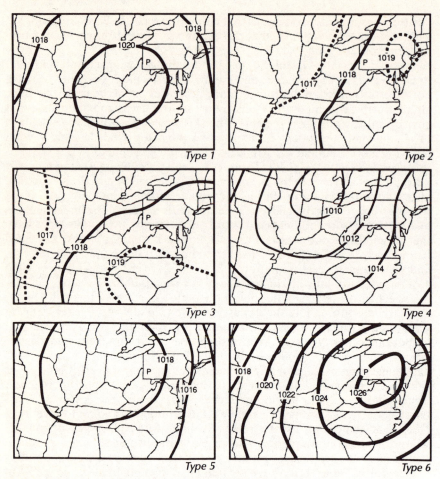

Figure 4.4 Eigenvector-based synoptic-type classification: average sea-level pressure pattern (mb) associated with the synoptic types.

frequent around the winter solstice than it is in midsummer, the average temperature is quite high. This denotes strong advection of maritime-tropical air into the northeast.

Type 8 is by far the coldest of all types, depicting a thick dome of polar or Arctic air over the midwest. Over Pittsburgh this means strong westerly winds and heavy stratocumulus clouds generated over the Great Lakes. Dry, cold air infiltrates the area, causing temperature and dew-point temperature to fall and visibility to increase beneath the cloud deck. This combination of conditions almost never occurs in summer but is observed on 6% to 8% of October-to-May days.

Type 9 shows well-defined high pressure directly east of Pittsburgh; Type 10 depicts a strong high west of the city. Therefore, they have fairly predictable, opposing tendencies. Pressure is high but falling with 9, while it

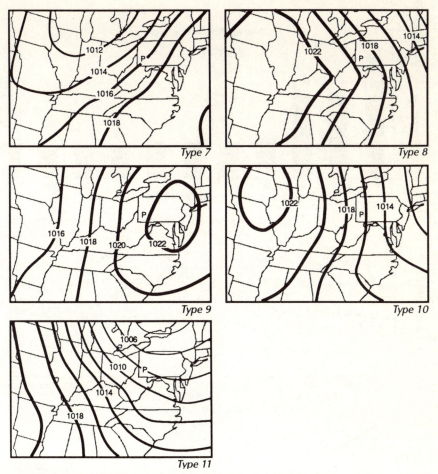

Figure 4.4 cont. Eigenvector-based synoptic-type classification: average sea-level pressure pattern (mb) associated with the synoptic types.

is lower but rising sharply with 10. Wind is light from the east-southeast on Type-9 days and strong from the northwest under Type-10 conditions. Trends in temperature, dew-point temperature, cloud cover, and visibility are also inversely related. Unlike the other nine types, neither Type 9 nor Type 10 have a strong seasonal preference.

Type 11 marks the passage of an intense wintertime low-pressure system to Pittsburgh's north. The lowest pressure and highest windspeed are measured on these days. The low and its associated fronts have just passed as the day begins. Pressure rises steadily throughout the day, while air temperature and dew-point temperature fall precipitously. Visibility starts to improve with the passage of the front but falls as the day goes on. Despite heavy lake-effect cloud, the total cloud cover decreases as the storm system pulls east. Although virtually non-existent during the warm months, Type 11

97

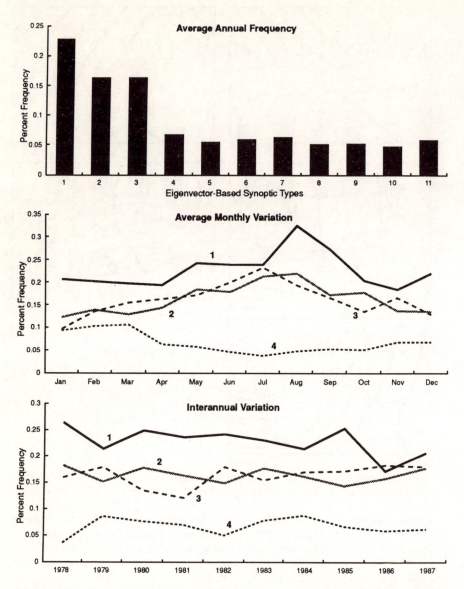

Figure 4.5 Average annual frequency of the eigenvector-based synoptic types (top), and average monthly variation (middle) and interannual variation (bottom) of the first four synoptic types.

happens on 12% to 14% of all December and January days.

On average, eight of the synoptic types appear only 5% of the time (Figure 4.5, top). As suggested above, this can be misleading because most types have a strong seasonal signal. For instance, in their season of peak occurrence, Types 5 (summer), 7 (winter) and 11 (winter) are observed on 11% to 14% of the days. In contrast, in their low-frequency season, they turn up on 3% of the days or less. Thus, the eigenvector-based synoptic-typing scheme captures the seasonal cycle quite well. Unlike studies using data that have not been deseasoned, however, all types appear in all seasons (for example, Figure 4.5, middle).

The synoptic-type classification also displays considerable interannual variability (Figure 4.5, bottom). For instance, Type 1, which occurred 22.7% of the time during the 10-year study period, varied from a high of over 26.3% (1978) to a low of 16.9% (1986). In that later year, Type 3, which had an average frequency of 16.2%, occurred on 18.0% of the days, surpassing Type 1. The sharp differences among the synoptic-type characteristics, plus the season-to-season and year-to-year changes manifested by the synoptic-type frequencies, suggest that the classification may relate reasonably well to variations in western Pennsylvania's surface environment.

A worked example of eigenvector-based map-pattern classification

Procedure description

The investigators followed the procedure for eigenvector-based map-pattern classification shown in Figure 4.1. These successive actions were taken and decisions were made.

— They selected the same 3603-grid data set prepared for the correlation-based classification of Chapter 3.
— Because this classification strategy requires S-mode decomposition, the team created a 35-by-3,603 data matrix. This represents the 35 grid points and the 3,603 days with 1200Z grids during the 10-year study period.
— To prevent the principal components from being swamped by the seasonal signal, the investigators deseasoned the data using the 13-day running-mean filter described in the eigenvector-based synoptic-type section of this chapter. The difference here was that departures from the 13-day map mean were calculated at each grid point. Thus, the spatial pattern on each map was preserved for the PCA.
— The investigators specified a correlation matrix.
— The team entered the data matrix into an orthogonally rotated PCA, which produced 35 components. To confirm that unrotated components would produce Buell patterns and that the orthogonally rotated components would produce credible modes of variation in the surface-pressure field, the team plotted components-loadings maps (Figure 4.6).
— Using (1) only those eigenvalues greater than one, (2) only those components explaining more than 5% of the variance, and (3) a scree plot of the eigenvalues to identify likely break points, the investigative team retained the first five principal components, which explained 94% of the variance in the pressure data.

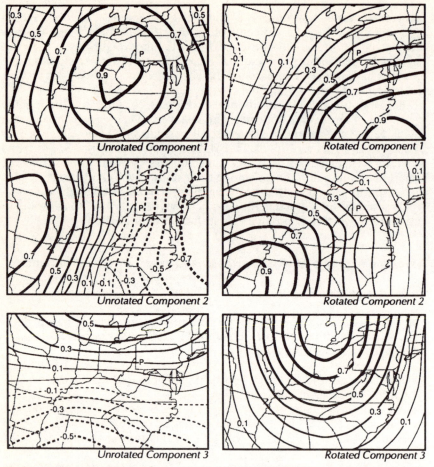

Figure 4.6 Components-loadings maps of the unrotated and rotated S-mode principal-components analyses. Note the classic Buell sequence in the unrotated components-loadings maps.

— The investigators applied Ward's clustering method to the scores of the retained components.
— Based on (1) a scree plot of the R^2 metric and (2) the number of types in the manual, correlation-based, and eigenvector-based synoptic-type classifications, the team decided that 12 clusters (that is, map patterns) were appropriate.
— To aid their understanding of the map patterns, the investigative team calculated the average pressure patterns for each of the 12 clusters.

Classification results

The mapped loadings of the first five unrotated principal components

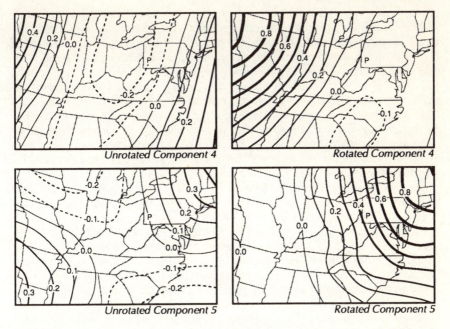

Figure 4.6 cont. Components-loadings maps of the unrotated and rotated S-mode principal-components analyses. Note the classic Buell sequence in the unrotated components-loadings maps.

display the classic Buell sequence (Figure 4.6). As predicted, the first component shows a bull's eye in the center of the study area. Unrotated Component 2 has an east–west opposition of loadings, while north and south are opposed in Component 3. Unrotated Component 4 displays a plus-minus-plus sequence from east to west. Component 5 has positive loadings in the southwestern and northeastern map quadrants; the northwestern and southeastern quadrants are negative. These patterns have few real-life analogs. The patterns of the unrotated loadings do not illustrate the main modes of spatial variation in the surface-pressure field but represent an artifact of the mathematical technique.

In contrast, the first five rotated principle components exhibit pressure patterns that are recognizable to the synoptic climatologist (Figure 4.6). Remembering that the loadings field must be interpreted in both its positive and negative phases, rotated Component 1 can represent either a coastal storm moving up the East Coast in winter (a Nor'easter) or, more likely, the Bermuda high in summer. One phase of Component 2 shows a storm heading from the lee of the southern Rocky Mountains to the Ohio River Valley; the other displays a high over the same area. Component 3 indicates either a storm centered over the Great Lakes, or a high over the area. Component 4 suggests an advancing Alberta low, displacing a weak Bermuda high; alternatively, an outbreak of cold polar or Arctic air could be ousting a weak low-pressure system over the southeast. Finally, the fifth rotated component denotes a Nor'easter that has tracked up the East Coast, an Alberta clipper that has passed over the Great Lakes into New England, or a Colorado low that took a slightly more southerly route than normal. In contrast, it could illustrate the appearance of a strong high over New England and the Canadian Maritimes. In summary, the rotated component loading patterns not only represent the main modes of variation in the surface-pressure field, they portray physically reasonable pressure patterns. The most common pressure pattern over the region is the summertime Bermuda high. Commonplace storm patterns are Colorado lows, low-pressure systems over the Great Lakes, and Alberta clippers. Polar- and Arctic-air outbreaks are frequent, and occluding cyclones typically exit the region in the vicinity of the St. Lawrence River Valley.

Few daily pressure maps, however, display these statistically purified patterns. Instead, they often show a mixture of these modes of pressure variation. The weights of the component scores vary from day to day and reveal this scrambling of modes. Therefore, investigators apply a clustering algorithm to identify the most common groupings of component scores. These clusters cannot be captured in map form, but average pressure patterns (map patterns) can be calculated for each of the clusters. Figure 4.7 displays these composite pressure patterns for the 12 clusters determined by the investigative team, while Figure 4.8 contains information on the map patterns' average annual frequencies, monthly variation, and interannual variability. A brief discussion of each map pattern's characteristics follows.

— Map Pattern 1 shows the Bermuda high. It is the most common pressure pattern, appearing on over 19% of the days in a typical year. Although the Bermuda high is recognized in all seasons, it is primarily a summer pattern, peaking in July when it occurs 37% of the time. Map Pattern 1 is surprisingly regular in its occurrence, varying from a low of 16.4% (1979) to a high of 21% (1981).

— Low-pressure systems occur over the Ohio River Valley on 10% of all days. Map Pattern 2 is most frequent in the low-sun half year, cresting in February, when it is three times more frequent than in mid-summer. Like Map Pattern 1, Map Pattern 2 pressure patterns show little interannual variability.

— The second-most-frequent pressure configuration is Map Pattern 3: a high centered over the Great Lakes. It is most common from May to October, occurring on over one-quarter of all days in the last half of summer. It varies considerably from year to year. For instance, in 1978 it was observed on nearly 18% of all days, but the next year it appeared only about 11% of the time — a difference of over 60%.

— Map Pattern 4 shows high pressure in the upper midwest and suggests an outbreak of polar air on the day following the pattern. It is common throughout the year and showed little variation throughout the 10-year study period.

— Low pressure over the Great Lakes occurs on Map Pattern 5 days. It is primarily a cool-season phenomenon with peak frequencies between November and April, when it appears on about 7% of all days. Its

numbers vary little from year to year.
— Map Pattern 6 is a variant of the Bermuda high, occurring on about 10% of all days. The high-pressure cell is weaker than in Map Pattern 1 and a weak low-pressure field is seen over the eastern Great Lakes through New England. This pattern has a very strong three-part seasonal cycle, happening on about 15% of the days between May and August, 10% from September to December, and about 6% from January to April. It has only moderate interannual variability.
— The seventh map pattern places a well-defined high over the Mississippi River Valley about 7% of the time. It has a pronounced seasonal cycle, appearing on 14% of all January days but on only about 1% of the days in August. Thus, Map Pattern 7 represents a deep intrusion of polar or Arctic air into the eastern United States. As might be expected, its frequency varies dramatically from year to year. In 1981, over 9% of all days were Map Pattern 7, but the following year, fewer than half as many cold intrusions were witnessed.
— Map Pattern 8 positions a massive elongated ridge over the Appalachian Mountains. It can occur at any time but is most common in the autumn, when it is observed on approximately 9% of all days. In terms of its annual percentage, it is the most highly variable of the map patterns. Although on average it appeared on 6.5% of the study period's days, Map Pattern 8 occurred 2.7% percent of the time in 1983, and 10.3% in 1984. This denotes an increase of 380% in one calendar year.
— Like Map Pattern 7, the ninth map pattern shows a broad, cold high-pressure feature over the mapped area during the winter months. Unlike the former, Map Pattern 9 covers a much larger area and has a considerably higher central pressure. This suggests entrenched polar or Arctic air over the eastern two thirds of the United States. Map Pattern 9 is present on about 10% of all days on average, but this varies from 11 to 13% in the low-sun half year to about 4% from May to August. Interannual variations are significant, with changes of 50% to 100% being the norm throughout the study period.
— Map Pattern 10 has low pressure over New England and a steep east–west pressure gradient over the majority of the grid. Clearly, a cold front has just passed through Pennsylvania and drapes down the East Coast. This storm pattern is virtually nonexistent in summer and is somewhat common in winter. There is little interannual fluctuation.
— Map Pattern 11 places cold, high pressure over the Great Lakes and has a Colorado low moving out of the Great Plains. The pattern suggests that the developing storm will pass over Pennsylvania or just to its north. This is a springtime pattern, with peak frequencies in the period February though April; it does not occur in the summer and has relatively low numbers in the autumn and winter. Interannual variability is low.
— The 12th map pattern is much rarer than the others. Still, it is a significant weather maker for residents of the eastern United States. Map Pattern 12 shows intense low pressure off the Carolinas and, for the most part, represents Nor'easters. The small number recorded in September and

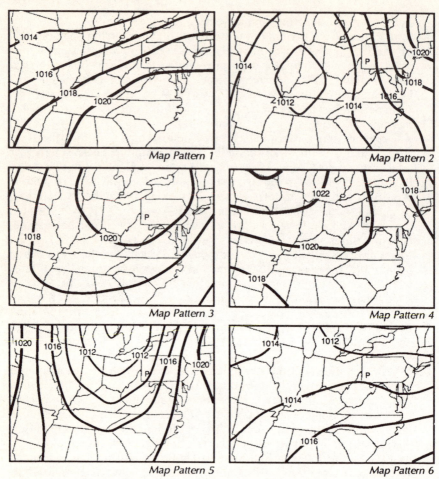

Figure 4.7 Eigenvector-based map-pattern classification: average sea-level pressure pattern (mb) associated with the synoptic types.

October primarily symbolizes tropical storms. Map Pattern 12 does not occur in the period May through August. Although year-to-year variation appears to be minor, varying from nearly zero to about 3%, this represents the difference between one and 10 severe storms rocking the East Coast. In terms of coastal erosion, property damage, and loss of life, this is an important difference.

The map patterns each bring relatively different climatic conditions to western Pennsylvania. Map Pattern 1 funnels warm, humid air into the region around the Bermuda high's western border, while Map Pattern 2 does the same in the early hours of the day. Later in the day, Map Pattern 2's low-pressure center directs a front over the study area. Map Patterns 3 and 4 tend to bring fine, cool-to-cold weather to the area, although Map

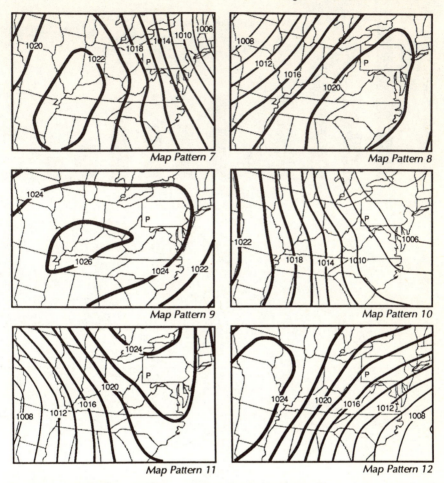

Figure 4.7 cont. Eigenvector-based map-pattern classification: average sea-level pressure pattern (mb) associated with the synoptic types.

Pattern 4 does have north-to-northwesterly breezes and can usher in lake-effect cloud and precipitation. The fifth map pattern is preceded by southerly advection of warm, moist air, which is quickly replaced by disturbed, stormy conditions. Map Pattern 6 brings relatively tranquil conditions with westerly flows into the region; the sky is probably hazy with mixed cloud but little precipitation. Lake-effect cloud and snow accompany Map Pattern 7. Strong northwesterly winds and low temperatures make these days particularly raw and unpleasant. Map Pattern 9 produces a much gentler winter day, with cold, crisp conditions and little wind; lake-effect clouds are likely but are probably restricted to the vicinity of the lake. Map Pattern 8, in contrast, produces warm, stagnant conditions. In summer it is associated with searing heat and poor visibility, while in winter these days are relatively warm and pleasant. Passage of a Map Pattern 10 storm

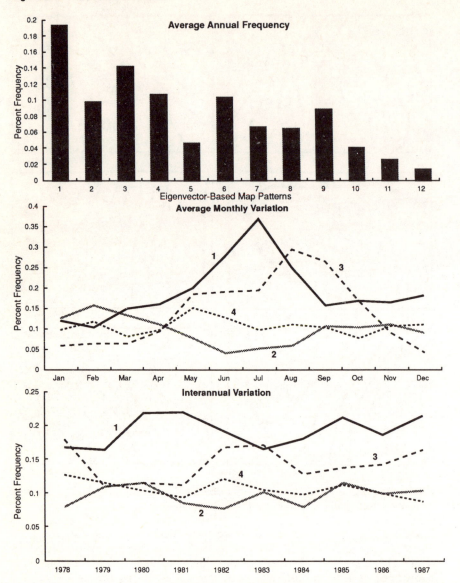

Figure 4.8 Average annual frequency of the eigenvector-based map-patterns (top), and average monthly variation (middle) and interannual variation (bottom) of the first four map patterns.

means intense northwesterly winds, lake-effect cloud and perhaps precipitation, but good visibility below the cloud deck or in those areas free of cloud. Map Pattern 11 days start out with gentle easterly winds. Over the course of the day, winds shift to southeast, then south, and finally southwest, advecting relatively warm, humid air into the area. These days are followed by strong storms which can dump significant quantities of rain or snow on the study area. The Nor'easters associated with Map Pattern 12 tend to deposit the bulk of their precipitation to the east of the study area. The Map Pattern 12 tropical storms are infrequent but can be significant events. For example, 1972's Hurricane Agnes dropped more than 250 mm of rainfall over western and central Pennsylvania.

This worked example suggests that eigenvector-based map-pattern classification is a potentially important tool in synoptic climatology. The average pressure maps depict those patterns the investigative team expected to see more than any of the other synoptic classifications presented in this book. Similarly, the annual frequencies matched the expectations of the investigators more than those of the other classifications. The mixtures of weak and strong seasonal cycles and of minor and major interannual variations suggest that the classification may capture atmospheric fluctuations important to the surface environment. In Chapter 6, I demonstrate that, indeed, the performance of the eigenvector-based map-pattern scheme is competitive with the other classifications.

5 Compositing, indexing and specification

Synoptic climatology consists of a diverse set of classification methods. In this chapter, I present three unlike methods: compositing, indexing, and specification. Composites are simply "average maps" of specific situations. On the one hand, synoptic climatologists like to use this environment-to-circulation method as a "first cut" at understanding the data. On the other hand, they resort to composites because various circulation-to-environment attempts at classifying the data have failed.

Indexing is a way to characterize variations in the atmospheric circulation with a simple time series. In one number, circulation indices can capture much of the richness of the more elaborate classification techniques. However, the depth of meaning conveyed by an index depends on its empirical and theoretical underpinnings; a superficial understanding of the atmospheric circulation results in a shallow index. Both compositing and indexing can be powerful tools if used with understanding.

Specification is a technique used more for prediction than analysis. Its roots are in long-range weather forecasting, and new versions are being developed for projecting the surface climates associated with global warming. None the less, it is a synoptic-climatological technique because, by definition, it specifies the surface climate from knowledge of the atmospheric circulation.

Both approaches to synoptic climatology (Figure 1.1) are represented in this chapter. Compositing applies an environment-to-classification approach to synoptic climatology, indexing employs the classification-to-environment perspective, and specification utilizes both. Accordingly, the model-comparison statistics of Chapter 6 cannot be used with composites but are appropriate for circulation indices and specification.

There are sections in this chapter on compositing, indexing, and specification. For each, I consider the methodologies and review the literature. I also present worked examples of compositing and indexing, which I use in Chapter 6 to evaluate the environmental scenarios. I do not offer a worked example of specification.

Compositing

Methodology

Compositing is easy to conceptualize and apply. Simply, the investigator selects a number of maps that satisfies an important criterion and averages them. As a form of environment-to-circulation classification, the classification criteria are independent of the atmospheric circulation, so there is no a priori choice of synoptic types prejudicing the results (Winkler,

1988). Say the goal of a study is to specify the typical surface-pressure pattern associated with exceptionally high daily precipitation totals at a single station. The investigator might decide that "exceptionally high" means equal to or greater than 50 mm per day. Then he or she would identify all days from the station record that meet this criterion, remove all daily grids from a digitized surface-pressure data set for those high-rainfall days, and calculate a mean map.

Investigators can introduce countless variations on this theme. Here are a few options.

— Complex criteria are possible. If the aim of a study is to determine the typical atmospheric configuration associated with acute acid-rain exposures, the investigator might want to composite pressure surfaces with exceptionally high-precipitation days *and* high-sulfate totals.
— Multiple composites can be made. If the acid precipitation at a station tends to occur as either low-precipitation total, high-concentration events or high-precipitation total, low-concentration events, then it might be appropriate to make two composites, one for each set of criteria.
— Composites are not restricted to pressure surfaces. If the research goal is to establish the average storm track associated with exceptionally high-precipitation events, the investigator can digitize storm tracks and average their positions.
— Investigators need not be tied to map surfaces. If an eigenvector-based synoptic-typing scheme is available for a station and the investigator wishes to know the air mass typically associated with the most acidic precipitation, he or she can calculate composite clusters.

In summary, compositing is flexible and amenable to almost any research problem.

Compositing can be the method of first and last resort. It is often useful to take a first cut at the data with compositing. For example, suppose an investigator suspects that the atmospheric circulation might control a crucial aspect of the surface environment. To elucidate this relationship, it is planned to propose a full synoptic climatology to a government funding agency. However, before committing resources, the agency wants to see strong evidence that this relationship exists. Applying theory and a simple research hypothesis (for example, there is a common atmospheric configuration on days when the critical variable exceeds some threshold), in a relatively short time the investigator can tease basic relationships out of the data set by compositing.

Sometimes investigators are sure that the atmospheric circulation must control the variations in a surface environmental variable. However, circulation-to-environment classifications, such as map-pattern classifications or eigenvector-based synoptic typing, fail to reveal a significant relationship. There are innumerable reasons why the association is weak: for example, the measurement of surface and atmospheric variables might be at incompatible time steps or there are too many missing measurements in the surface data. Environment-to-circulation approaches, such as composites, are often insensitive to these sorts of problems because of the way they are calculated. Thus, they often serve as the method of last resort. In Chapter 6, I present an example in which other synoptic classifications fail to explain surface environmental variability, but compositing successfully portrays this relationship.

Despite all of its positive attributes, there are at least three problems with compositing. First, composites can average disparate atmospheric settings, thus inducing a false impression of the typical situation associated with a given criterion. The investigator should always study the maps or other data used to create the composites. Standard practice should be to calculate and plot standard deviations when producing the composites.

Second, composites are no better than the criteria upon which they are founded. If the investigator produces composites based on false assumptions, poor theory, mixed populations, etc., the composites will portray erroneous information and relationships.

Third, because compositing is an environment-to-circulation approach to synoptic climatology (Figure 1.1), model-comparison statistics cannot be used with composites (Chapter 6). Composites of the atmospheric circulation are neither independent of the surface environment, nor are they continuous in time: two requirements necessary for calculating model-comparison statistics.

In summary, although it has its flaws, compositing is an indispensable instrument in the tool kit of the synoptic climatologist. Compositing's simplicity and flexibility enable it to complement and, in some instances, replace the other more complicated but less flexible synoptic-climatological tools. In the next two sections, I review the compositing literature and present a worked example of compositing, respectively.

Previous studies

The literature abounds with composites. However, surprisingly few studies employ compositing as the sole synoptic-climatological technique. For example, Tyson (1981), Harrington and Harman (1985), Yarnal and Diaz (1986), and Yarnal and Leathers (1988) use composites with indexing. I review their work in the discussion of indexing. Here, I present a few major studies in which compositing is the primary method, with one exception. In that case, I discuss one investigation in which compositing is nested with other synoptic-climatological techniques to reveal the power of this strategy. In all cases, I emphasize the findings and not the method, thereby illustrating the range of compositing studies tackled by various investigators.

Comparisons of composites based on contrasting criteria can reveal a great deal about a surface environment. To determine synoptic controls on glaciers in the Brooks, Alaska, and Coast mountain ranges of Alaska, Fahl (1975) constructed composite surface-pressure maps and surface-pressure anomaly maps for summer snowfall, winter snowfall, and summer hot spells. He compared them to average conditions; that is, composites of all maps.

He found, in general, ridges and anticyclonic curvature produce hot spells, while troughs and cyclonic curvature produce snow. In each range, glacier growth can only occur if moisture is available; if the circulation regime cuts off the moisture source, strong negative mass balances result in rapid glacier retreat. Because of the unique geography of Alaska, each range's glacier behavior is independent of the others. The situation is most complex in the Brooks Range because of many interacting factors such as sea ice and alternating northerly, marine and southerly, continental air flows. The Coastal Range, where mass balance is controlled by synoptic systems coming off the Pacific Ocean, is least complex.

Although Achtor and Horn (1986) suggest that composites should be based on events rather than time intervals, investigators can define their compositing criteria as periods of time. For example, Vukovich and Fishman (1986) constructed July and August composites because these are the two periods of highest O_3 and SO_2 concentrations in the United States. They plotted the following maps for the eastern two thirds of the country: mean monthly surface pressure; mean monthly O_3; mean monthly SO_2; and anticyclone paths for July and August. Their pollution data set represents the large-scale background concentrations unencumbered by local-scale variations. They found that the distribution of O_3 (a secondary pollutant) is significantly more variable than SO_2 (a primary pollutant) from year to year. SO_2 concentrates in the Ohio River Valley, a principal source region. Since the conversion of SO_2 to sulfate is on the order of one day, this strong relationship is not surprising. There is no apparent correlation between the distribution of mean pressure and O_3 and SO_2. However, high O_3 means locate in the tracks of migratory anticyclones, especially those that stagnate. They also found that cities are not the only source regions for ozone; significant source effects are observed in urban and upwind areas.

Winkler (1988) produced integrated composites for forecasting heavy precipitation in summer. Five years of extreme precipitation events (≥ 0.75 mm per 24 hours) netted 43 such rainstorms over Minnesota. She broke the storms into four types, based upon their areal extent, the number and orientation of the rain cells, and the ratio of the rainfall pattern's major-to-minor axis. Each type occurs with equal frequency. For each of the rainfall categories, she computed the following composites: surface pressure, temperature, and dew point; 850-mb height, temperature, dew point and wind speed; 700-mb height, temperature and dew point; 500-mb height and temperature; and 300-mb height and wind speed. She constructed composites for the day of the storm (t) and for the days immediately preceding and following the event ($t-1$ and $t+1$, respectively). Finally, she manually consolidated each type's important circulation features from the different levels onto one map. These maps are *integrated composites* or, as she called them, *synoptic analog models*. First Winkler studied the time of day these heavy-precipitation events occur, the duration of the events, and the likely flash-flood danger from each event. She then conducted a detailed analysis of each type's integrated composite at $t-1$, t and $t+1$, focusing on their circulation characteristics. The results show that the synoptic-scale circulation regime determines the spatial characteristics of extreme rainstorms: that is, where the heaviest precipitation will fall. Forecasters can use knowledge of this relationship to improve subjective forecasts of location and areal extent of extreme rainfall events.

Farmer *et al.* (1989) considered the relationships between large-scale composite pressure-anomaly patterns and sulfate concentrations in precipitation (that is, acid rain) at Eskdalemuir, Scotland. They constructed two types of composites: one using actual surface-pressure values and another based on point-by-point anomalies. They conducted t-tests on the anomaly maps and established that the statistically significant points formed spatially coherent and discernible regions, thus highlighting the main features of the anomaly fields. Their findings show that the months with the highest sulfate concentrations, which only occur from February to September, tend to associate with westerly flow, weakened gradients, and a hint of blocking over the mid-Atlantic, all of which translates into increased southerly and easterly flow over Scotland. Thus, the individual synoptic events, which comprise these monthly pressure data, can transport more sulfate to Eskdalemuir from the SO_2-producing regions of England and the continent. During the low-sulfate months February to September, the flow is much stronger, with strong gradients, and westerly-to-southwesterly flow. Thus, synoptic systems and clean air from the Atlantic are sweeping over the site.

As noted above, composites can be powerful when linked to other synoptic-climatological methods. Using satellite imagery, Carleton (1986) manually identified bursts and breaks in the southwest United States' monsoon for three summer seasons. He constructed composite maps of surface pressure and 500-mb heights for each of the three summers and for all three summers combined. He also studied maps for the day before and day after the peak of each burst. From this he developed a manual synoptic-type classification of eight 500-mb patterns associated with bursts and breaks (see also Carleton, 1987). The composites of surface pressure show little difference between bursts and breaks. However, the 500-mb differences are great, and it is clear that most bursts associate with troughing, while all breaks associate with ridging. Apparently, westerly cold-core disturbances invading the tropics destabilize the atmosphere. Interannual differences in the mean position of the troughs and ridges are evident. Nevertheless, the 500-mb types indicate that not all monsoon bursts are the result of the southward penetration of westerly troughs. Anticyclonic conditions are also important, and relatively small shifts in latitude of the Bermuda and North Pacific highs determine whether burst or break conditions will prevail.

Worked example

Procedure description. The investigative team had to determine two things: (1) what they wanted to composite and (2) the surface-based criteria for selecting the synoptic data. For the former, they aimed to calculate a four-day sequence of composites representing the typical evolution of a high-ozone event in Pittsburgh, Pennsylvania. To do this, they would have to calculate the average pressure pattern associated with high ozone, plus the pressure pattern two days before, one day before, and the day after an event.

Compositing uses an environment-to-circulation approach to synoptic climatology. Thus, the investigators needed to choose surface-based criteria most likely to reveal a pressure pattern characteristic of a high-ozone event. The best way to show well-defined patterns, if they exist, is to select a representative sample of extreme cases. The Environmental Protection Agency's National Ambient Air Quality Standard (NAAQS) for surface ozone is 120 ppb. During the 10-year study period, there were 30 days in which Pittsburgh's maximum one-hour ozone concentrations exceeded this standard. Consequently, the investigators culled the synoptic data for the four-day sequences around these 30 high-ozone events.

The synoptic data used were the same five-by-seven grids of surface pressure described in Chapter 3 and employed in the manual, correlation-based, and eigenvector-based map-pattern analyses. The investigators wrote a simple computer program to average the grids and entered the 30 grids for each day into the program. (See Chapter 6 for details of the surface-ozone data and the application of the other synoptic classifications to surface ozone.)

Results.   The composites paint a physically reasonable picture of a high-ozone event in Pittsburgh (Figure 5.1). Two days before the event, a huge high-pressure cell dominates the eastern half of the United States. Under this cell, the pressure is weak and wind is light, sunlight is strong, and temperature is high; all conditions favorable to ozone formation (Comrie, 1990). Polluted air from the industrial south is lazily transported into the midwest, while in Pittsburgh the precursors of ozone stay in place.

The day before the event, the anticyclone migrates somewhat to the east and the pressure gradient tightens a bit, encouraging slightly stronger southwesterly flow. Pittsburgh has been under the center of the cell for at least two days now. Sunny, warm conditions continue, supporting the photochemical buildup of ozone locally and over the Ohio River Valley, the most industrialized region of the United States.

On the day of the high-ozone event, the high pressure cell is still strong, but it is moving increasingly eastward. The area to the west with anticyclonic curvature has the classic back-of-high form. Wind is very light over Pittsburgh, but with eastward movement of the anticyclone, the wind will strengthen significantly and develop a southwest orientation. Pittsburgh, which has been stewing under locally produced ozone, is now adding to the pot by importing ozone from the southwest. Some of this additional ozone has been traveling for days from the south, but a larger proportion has been forming over the Ohio River Valley. The sky is still cloud free and temperature is high, although visibility has probably been decreasing throughout the event as smog has built up.

The day after the event, the composite pressure pattern is not as well defined. A high is present to the south-southeast of Pittsburgh and weak low

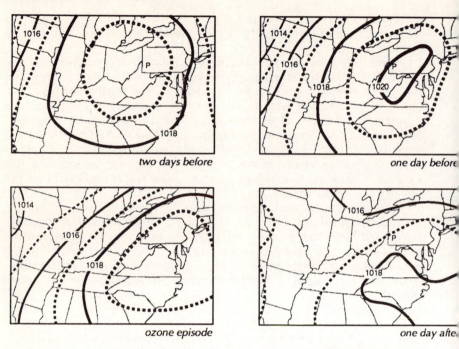

two days before

one day before

ozone episode

one day after

Figure 5.1 Surface-pressure composites (mb) for all days when Pittsburgh's daily O$_3$ maxima was greater than 120 ppb (modified from Comrie, 1990, with permission of Edward Arnold Publishers).

pressure has formed north of the Great Lakes. Air flow over Pittsburgh is southwesterly and the sky is probably cloudy. This composite suggests two related situations. First, on many days making up the composite, the anticyclone has persisted and the ozone level is high but probably below the benchmark of 120 ppb. Imported ozone is more important than local ozone because the cloud cover has stopped much of the local formation. Alternatively, in many cases, a cold front has moved through, or will soon be moving through the area. Until the front passes, the ozone level is high because of southwesterly transport, but strong wind and rain from the front will lower the ozone value significantly. There is little doubt that, in either case, the event is winding down and the second day after the peak will have a lower ozone concentration.

The investigative team also constructed composites for the other environmental scenarios. As is true of Pittsburgh ozone, each of these composites was an important interpretive tool, adding to the investigators' understanding of the circulation-to-environment classifications. Worked examples of some of these composites are presented in Chapter 6.

Indexing

Methodology

Like compositing, synoptic-climatological indexing is simple to conceptualize and use. An index consists of a series of values depicting the variations in the atmospheric circulation over time. After compiling an index, the investigator relates the time series of the circulation to one representing the surface environment, usually with a simple correlation coefficient. By relating the circulation to the surface environment, indexing fits the working definition of synoptic climatology used in this book. Because the index is determined independent of the surface environment, indexing is a circulation-to-environment classification.

Index values are a distillation of the atmospheric circulation to a single number. Computation of this number can be as elementary or as elaborate as the investigator wishes. For instance, a simple index of circulation strength might calculate the pressure difference between 60 °N latitude and 30 °N along one meridian. A complicated index might use 23 functions derived from linked eigenvector analyses.

Simple does not translate into *worse*; *complicated* does not mean *better*. Some of the most elegant circulation indices are surprisingly simple in their calculation. For example, the Pacific–North American (PNA) teleconnection index discussed in the next section uses just three normalized mid-tropospheric height anomalies (Yarnal and Diaz, 1986). Yet the PNA index not only captures the state of this one teleconnection, it conveys direct information on the zonality of North Pacific–North American circulation (Yarnal and Leathers, 1988). The index is based on the geographical anchoring of the global atmospheric circulation; that is, the PNA pattern owes its existence to the presence and nature of the Tibetan Plateau, Cordillera of North America, and North Pacific and North Atlantic basins. The PNA pattern further implies coupled relationships among Asian snow cover, the state of the ocean-atmosphere system of the tropical Indo-Pacific basin, the East Asian jet stream, North Pacific sea-surface temperature, and North American climate (Leathers, *et al.*, 1991; Leathers and Palecki, 1992). Thus, this deceptively simple index contains explicit and implicit information about global climate dynamics and the state of the Northern Hemisphere climate.

The potential represented by synoptic-climatological indices is considerable. Indices founded on extensive empirical and theoretical work, such as the PNA index, produce results that reflect the dimensions of that foundation. Conversely, indices that lack this underpinning are more superficial. This is not necessarily bad, if the goals of the research are limited in scope: for example, the investigator has a simple, one-dimensional research hypothesis or the project has a single applied environmental focus. No matter whether the synoptic index is rich or lean, it must be suited (if not tailored) to the environmental problem to which it is linked.

Previous studies

Indices have been applied to surface environments all around the world. Accordingly, I address synoptic indexing from a geographical perspective. From the range of articles reviewed, it will be clear that indexing is usually used to represent large synoptic-scale to global-scale circulation systems, rather than small to mid-range synoptic-scale configurations.

Several investigators have studied long-wave control over the climate of the 48 contiguous United States using synoptic indices. In one of the most recent works, Leathers *et al.* (1991) plotted correlations between monthly data from the 344 American climatic divisions and the PNA index. For temperature, they found that large areas of the southeast and northwest associate strongly with the teleconnection pattern, with variations in the index explaining more than 90% of the variance in some climatic divisions. They also observed that the area of strongest correlations has an annual cycle, suggesting that coupled north–south movements of the circumpolar vortex and east–west fluctuations of the main trough–ridge axes control surface-temperature variations. For precipitation, relationships are not as strong, although the PNA index explains a large proportion of the winter and spring precipitation variability over the Ohio River Valley. Leathers and Palecki (1992) went on to demonstrate the links between this long-wave control of American climate and variations in the global climatic system. In an earlier work on this problem, Diaz and Namias (1983) computed circulation indices by averaging 700-mb heights for huge sectors over the North Pacific Ocean and North America. They correlated these indices with seasonal average temperature and precipitation for nine different regions of the United States. In essence, Diaz and Namias found the same associations between the circulation and surface climate that Leathers *et al.* uncovered, except that the earlier study used much cruder spatio-temporal data and could not make the links to global-climate dynamics afforded by the PNA ·index. In a similar investigation, Skeeter and Parker (1985) calculated meridional and zonal indices for 500-mb flow over the United States and compared them to surface temperatures at 150 stations. To construct their indices, they computed 500-mb height differences along two east–west profiles to detect the strength of meridional flow aloft and along three north–south profiles to determine zonal-flow intensity. Again, their results are similar to Leathers *et al.*, although they have neither the spatio-temporal resolution of the more recent work, nor the linkages to global forcing mechanisms. Their results are an improvement, however, over the cruder findings of Diaz and Namias. Thus, these three studies show how a combination of finer surface data and the evolution of climate-dynamics theory enabled investigators to improve synoptic-climatological indices and the understanding of atmospheric controls on surface climate.

Synoptic indices have also been used to study the climates of various regions of the United States and Canada. In the arid southwestern United States, the most important determinant of summer-precipitation variability is the misnamed "monsoon." To determine the synoptic controls on the cloud cover associated with this feature, Carleton (1985) developed several indices from surface and 700-mb charts. These included weekly measures of: the intensity of the summertime heat low over the region; the intensity of the North Pacific high; the latitude of the Bermuda high; the intensity of the Bermuda high; the strength of the midlatitude westerlies; and moist, southerly flows into the arid southwest. He found that no single index explains all the variance in cloud cover. The index that relates best to monsoon cloudiness is the latitude of the Bermuda high, although the behavior of the North Pacific high strongly modulates this relationship. Turning to the western Great Lakes region, Harrington and Harman (1985) used a combination of manual map-pattern classification, compositing and indexing to identify synoptic controls on moisture stress. They distinguished two synoptic groups associated with distinct moisture stress: one with moisture-stress gradients trending southwest to northeast, and 500-mb flow from the northwest; and an another with an east–west gradient, with 500-mb flow from the southwest. The latter synoptic group is important in determining regional moisture stress in the late spring, while the former relates to moisture variations in the summer. In work on the West Coast of Canada and the United States, Yarnal and Diaz (1986) set out to determine relationships between regional climate and El Niño-Southern Oscillation (ENSO) events. They found that surface climate associated with ENSO varies wildly but through compositing identified a PNA-pattern signal. Using the PNA index, they demonstrated that the two extreme phases of ENSO (El Niño and La Nina) are related to strong PNA patterns (of opposite signs) about half of the time; the remaining half has flow regimes approximating the climatological mean. Their results suggest that investigations of North American climatic variability should focus on the PNA pattern and its causes, not ENSO. Yarnal and Leathers (1988) continued on this theme, studying Pennsylvania's climatic variability with a combination of composites and indexing. They demonstrated that interdecadal climatic variations in this region are associated with PNA-like variations, while interannual variations are tied to the PNA pattern and the North Atlantic Oscillation (NAO), another important teleconnection.

Moses et al. (1987) looked at the NAO and its relationship on surface climate around the North Atlantic basin. One phase of the NAO index relates to strong zonal flow, below-normal temperature in Greenland, and mild, moist conditions in Europe. The other phase has extreme meridional flow and blocking, with above-normal temperature in Greenland and below-normal temperature, dry conditions, and locally heavy snowfall in Europe. Like many investigators before them, they demonstrated that interannual climatic variability is closely tied to the NAO. However, they discovered that removing the effects of the NAO from the record does not explain long-term trends in North Atlantic–European climate.

Jacobeit (1987) studied the relationship between daily trough positions over the Mediterranean Sea and precipitation variability around the basin. He developed a number of 500-mb indices to compare to 101 Mediterranean precipitation time series, including mean latitude and longitude of troughs, intensity of the flow, geopotential height anomaly, relative vorticity, and others. Jacobeit incorporated upstream flow over the Atlantic Ocean because of its influence on the Mediterranean. He found that low-latitude zonal flow over the Atlantic reduces trough frequency over the Mediterranean, thus decreasing precipitation around the basin. Meridional flow and cyclonic anomalies over the eastern North Atlantic produce less troughing in the western Mediterranean and more troughing in the east, with increased precipitation along the eastern border. Meridional circulation with anticyclones over the eastern Atlantic relates to quasi-stationary troughing in the western Mediterranean and concomitant precipitation increases in the central Mediterranean.

Africa south of 22° experiences somewhat regular oscillations between nine-year wet and dry spells. Concurrent with this 18-year cycle, the south coastal region suffers a 10-year alternation of five-year runs of wet and dry climate. Tyson (1981) used indexing to explain the atmospheric controls on these two quasi-periodicities. He constructed 16 annual station-pair indices at 850-mb and another 16 at 500-mb to find correlations between the atmospheric circulation and rainfall departures. These indices served as a surrogate for geopotential-height grids, which were not available for this data-sparse area of the world. The results showed that latitudinal variations in the subtropical ridge of Wave 1 appear to force the quasi-periodic, 18-year oscillation in rainfall, while longitudinal shifts in the Indian Ocean ridge of Wave 3 seem to be responsible for the 10-year oscillation around the southern Cape. To determine more precisely the control of the atmospheric circulation over these regional variations, Tyson (1984) related his 1981 annual synoptic indices to rainfall data from 60 stations distributed evenly over the subcontinent. Correlating the time series of each level's 16 synoptic indices with the 60 precipitation stations called for data reduction, so he subjected the synoptic indices to an orthogonally rotated, S-mode PCA. The PCA produced four significant components (that is, a new set of indices) at 850 mb and six components at 500 mb. Tyson demonstrated that the atmospheric controls on wet and dry spells vary across South Africa. For instance, rainfall variations in central and northern interior parts of the country are caused primarily by changes in the tropical easterlies and easterly waves. Precipitation variations along the coasts, in contrast, are generally associated with the changing positions of the subtropical ridges.

Rogers (1983) used zonal and meridional indices to study relationships between the Southern Hemisphere's atmospheric circulation and the surface temperature in Antarctica. He related his indices to an eigenvector-based regionalization of surface temperature. The results were difficult to interpret because he based the regionalization on common factor analysis rather than PCA. In general, he concluded that strong westerlies mean mainland Antarctic temperature is lower than normal, while peninsular stations are anomalously warm. Conversely, weak westerlies relate to a relatively warm mainland and cold Antarctic Peninsula. Fluctuations in the westerlies explain one-quarter to one-third of the temperature variance. In the equinoctial seasons, these relationships were neither as clear nor as strong. In addition to showing the role of long waves on Antarctica's temperature regime, the results also point out, first, the seasonal variability of various circulation controls on surface temperature and, second, the interannual fluctuations in preferred Antarctic coastal locations for wave cyclones.

The most commonly used index in the literature is the Southern Oscillation Index (SOI). This index indicates the state of the coupled ENSO system of the Indo-Pacific basin. Although it is not expressly an index of the atmospheric circulation, it does imply the low-level and upper-level atmospheric flow over the tropics, as well as changes in the global-scale mass and energy fluxes driving those winds. The SOI is often used to determine teleconnections within the tropics (Yarnal and Kiladis, 1985) and between the tropics and the extratropics (Yarnal, 1985b). See these two references for early synoptic-climatological applications of the SOI, and Philander (1990) for more recent investigations.

In summary, indices are a simple way to capture the essence of large-scale circulation systems on time scales varying from days to decades and longer. Although there is no reason to believe indexing should perform poorly at smaller synoptic scales, few studies using this technique at smaller scales have been published. In the following, I present a worked example of an index based on just such a smaller scale.

Worked example

Procedure description. The investigative team developed a synoptic index to relate pressure-pattern intensity, and direction and sign of the pressure gradient to the surface environment. To determine the index, the investigators identified four NMC grid points forming a cell over western Pennsylvania. To make sure that the index measures pressure magnitude and gradient independent of the season, they deseasoned the pressure data using the 13-day running-mean filter described in Chapter 4.

In the early stages of index development, the team did not have a good idea of how pressure-pattern intensity and gradient relate to the surface environment. To refine their ideas, they regressed various combinations of pressure data and pressure-difference measures against the surface environmental data used in the scenarios of Chapter 6, such as Pittsburgh surface-ozone or wet-sulfate deposition at Penn State University. Finally, they settled on an arbitrary combination of the various regressions that produced the following formula:

$$0.6374 \, P - 0.636 \, \Delta P - 0.2538 \qquad (5.1)$$

P is the average deseasoned pressure at the grid cell's four points, and ΔP is the pressure difference between the two eastern grid points and two western grid points. The synoptic index resulting from this formula varies from 1.0 to -1.0.

The use of regression suggests that the surface environment controlled the selection of the pressure data, making the synoptic index an environment-to-circulation classification. By arbitrarily combining the regressions, the investigative team, however, determined equation 5.1 without intentionally

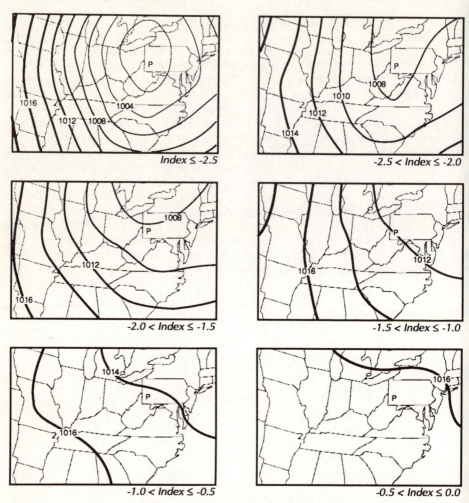

Figure 5.2 Synoptic indexing: average sea-level pressure pattern (mb) associated with the synoptic-index categories.

giving preference to any specific environmental variable. Furthermore, in Chapter 6, the index is applied to the environmental scenarios in the same way as the circulation-to-environment variables, that is, without isolating specific environmental criteria. Thus, like specification (see the following section), the index and its application fall somewhere between environment-to-circulation and circulation-to-environment classifications.

To develop a visual appreciation of how the synoptic index works, the investigative team constructed mean-pressure maps of index categories. To do this, they first divided the index into half-standard-deviation blocks and then composited the regional five-by-seven surface-pressure grids for all days that fell in those categories. For interpretive convenience, I display

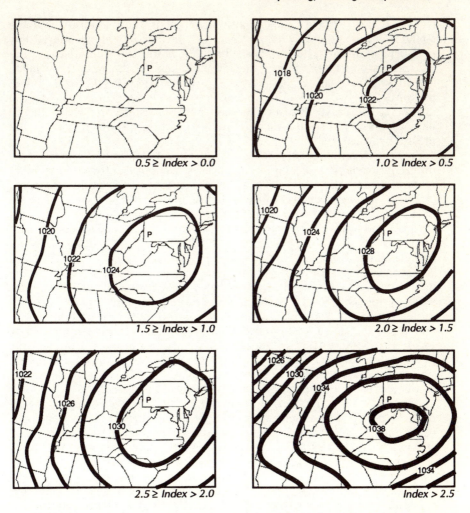

Figure 5.2 cont. Synoptic indexing: average sea-level pressure pattern (mb) associated with the synoptic-index categories.

these maps (Figure 5.2) and related graphs (Figures 5.3 and 5.4) using Z-scores expressed in standard-deviation units.

Results. The mean pressure patterns associated with the synoptic-index categories show significant variations in the low- and high-pressure fields over western Pennsylvania (Figure 5.2). On those days with index values of less than −2.5 standard deviations, intense low pressure is centered over the study area. This pressure pattern is reminiscent of a powerful storm tracking up the Ohio River Valley, with annual average central pressures below 1,000 mb. The mean pressure patterns for Z-score values from −2.5 to −2.0 and from −1.5 to −2.0 are only slightly different in their intensity, but the configurations are somewhat distinct. The former has more focused low pressure, while the latter displays a broader, flatter low. In both cases, the low pressure is focused over the Great Lakes. In the next three index categories, that is, −1.5 to −1.0, −1.0 to −0.5, and −0.5 to 0, lowest pressures are over New England and the Canadian Maritimes, or even farther to the east. Pressure gradients over western Pennsylvania are increasingly weak, so that the −0.5 to 0.0 composite shows a virtually flat pressure surface.

From a climatological perspective, positive index values show a very different set of pressure patterns. The first category, 0.0 to 0.5, shows a featureless pressure surface. The pressure surfaces representing Z-scores of 0.5 to 1.0, however, display a distinct high centered over Virginia. In fact, all of the subsequent pressure fields depict increasing high pressure focused on Virginia and not western Pennsylvania. Furthermore, each successive map reveals increasingly steeper pressure gradients over the midwest but not western Pennsylvania.

Graphical plots of the synoptic index reveal a great deal about the atmospheric environment over western Pennsylvania. Over the 10-year study period, the synoptic index has a periodic variation (Figure 5.3). This periodicity is not sinusoidal but is regular in the absolute variability above and below the zero line. Summers vary little, while winters are marked by great positive and negative departures in the pressure field. Summertime Z-scores greater than 1.0 are infrequent, and departures exceeding 1.5 standard deviations are rare. Thus, even with removal of the seasonal cycle, winter and summer pressure-pattern intensities are distinct.

Another interesting fact brought out by the long-term record is the difference in the frequencies of intense highs and lows. Strong highs (Z-scores > 2.5) occurred only 10 times during the study period, whereas 33 strong lows (Z-scores ≤ −2.5) were observed. Only 2 highs had Z-scores greater than 3.0, while 19 lows had values less than or equal to −3.0. Therefore, intense highs are much less frequent in this region than strong low-pressure systems. Alternatively, these index figures could indicate that the atmosphere places strong upper bounds on high-pressure systems and weaker lower limits on storms.

Although the points noted above remain true, the synoptic index possesses considerable interannual variation in the regional surface-pressure field (Figure 5.4). For instance, comparing the first two years of the study period, the following are clear:

— Late winter and early spring 1979 had many strong storms, while the corresponding period for 1978 had few.
— The 1978 index shows only a few low-pressure systems (values ≤ −1.0) passed over the area from late spring until late November. Even systems with Z-scores ≤ −0.5 were somewhat infrequent throughout this period. In contrast, the next year had more than a dozen systems with Z-scores ≤ −1.0 during the same period, and dozens of low-pressure fields with values ≤ −0.5.

Figure 5.3 Variation of the synoptic index throughout the study period. Missing values are set to zero for illustrative purposes.

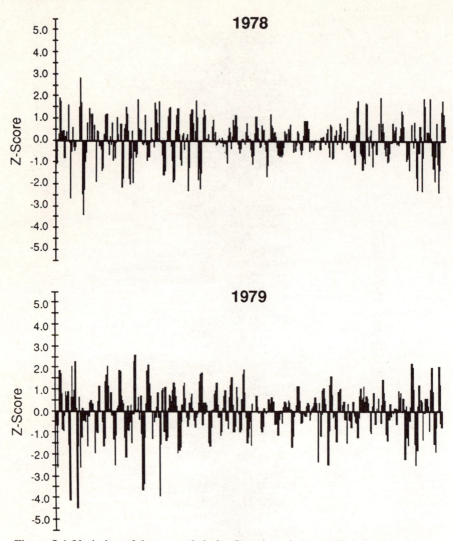

Figure 5.4 Variation of the synoptic index for selected years. Missing values are set to zero for illustrative purposes.

— The tranquil summer period was several months long in 1978 but lasted only a month or two in 1979.

In summary, the synoptic index suggests that 1978 was a relatively calm year over western Pennsylvania with a long summer. Conversely, 1979 was a disturbed year, with intense winter storms, a longer, unsettled spring, a short summer, and an early, stormy autumn. Study of the other years' synoptic indices reveals similar intriguing information. For instance, 1982 was marked by a long, gentle spring and summer, while all of 1986 exhibited

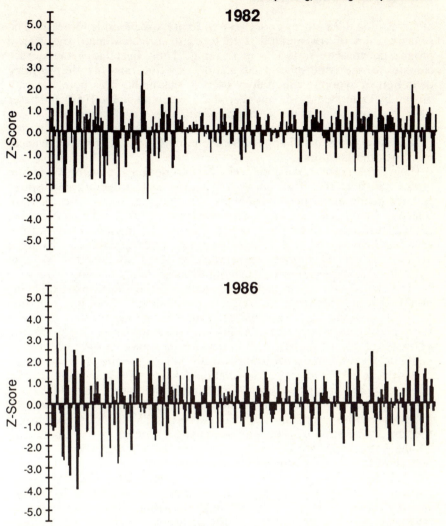

Figure 5.4 cont. Variation of the synoptic index for selected years. Missing values are set to zero for illustrative purposes.

a roller-coaster ride of stormy weather punctuated by intense high pressure.

The synoptic index provides a unique perspective on the atmospheric circulation over western Pennsylvania. I address the question of how it relates to the surface environment, especially in comparison to the other classifications, in Chapter 6.

Specification

Specification is a statistical procedure originally designed by the United

States National Weather Service (NWS) to prepare monthly forecasts. The principal aim of the method is to forecast surface climate by using a prognostic pressure-surface map as input. Thus, by relating the surface climate to the atmospheric circulation, specification fits the working definition of synoptic climatology used in this book. This point was not missed by W.H. Klein, who pioneered the technique in the 1950s and bases his consulting service on it today. Klein uses specification to prepare synoptic climatologies, as well as forecasts. In this section, I describe the technique as devised and applied by Klein, noting variations and advances made by other authors.

Klein (1983) provided a good technical description of how the procedure works and how the NWS employs it to forecast temperature. Gridded 700-mb geopotential-height anomalies are fed into a stepwise forward-selection (that is, multiple regression) analysis one point at a time to predict temperature at one surface station. For example, Klein used a 133-point grid covering the northeast North Pacific, North America, and the western North Atlantic to predict surface temperature in the United States. He also entered the surface station's mean monthly temperature into the regression, for a total of 134 terms. Klein devoted much of the 1983 article to explaining how he determined the number of predictors to include in the final equation for each station. In Klein (1985a), however, he demonstrated that a shortened equation, selected on a subjective basis and containing an average of five terms, tested as well or better than longer equations selected by other methods. He predicted temperatures at 109 American cities in this manner.

Besides methodological considerations, Klein (1983 and 1985a) also discussed relationships between the surface temperature and the 700-mb height anomaly fields. In general, he found that the procedure selects the previous month's temperature as an important predictor about two-thirds of the time, suggesting considerable persistence. One to two of the height-anomaly predictors are always local, suggesting a barotropic relationship between the atmosphere and surface temperature. The remaining height-anomaly predictors are remote and almost always to the west of the station, implying that westerly flow and eastward propagation of barotropic teleconnection centers are crucial in determining surface temperature. Averaged over the United States, the surface-temperature specifications explain more than 70% of the variance from December to June, but only about 65% from July to November. Klein's other principal conclusions were:

— Local boundary-layer phenomena, such as sea breezes, snow cover, fog and low clouds, reduce the accuracy of the specification.
— The standard error of estimate and standard deviation of temperature are largest in northern and central regions during winter and smallest in southern and coastal regions in summer.
— Month-to-month temperature persistence and the contribution of the previous month's temperature are largest in coastal regions and in midsummer and smallest inland and in fall.
— Geopotential height anomalies to the west of the station are more

important than height anomalies to the east or local height anomalies in specifying surface temperature, especially in winter and spring.
— Height anomalies in only a few key areas (for example, the PNA-index points of Leathers *et al.*, 1991) explain most of the temperature variability during winter and spring; the critical areas are not as concentrated or as well defined in summer and fall.

Klein (1985b) and Klein *et al.* (1989) applied specification to Alaska and Canada, while Klein and Yang (1986) employed the technique with European data. Each study made similar overall conclusions, but with regional variations and varying degrees of success.

Using the same data as Klein (1983 and 1985a), Klein and Kline (1984) and Kline and Klein (1986) utilized specification to determine the synoptic climatology of the United States in winter and summer, respectively. If regression equations are considered indices, then the Klein and Kline articles are really compositing and indexing studies, and therefore very much in the mainstream of synoptic climatology. In winter, they found that large-scale correlation patterns produce a PNA-like pattern for daily, five-day, and monthly mean data. Moreover, the magnitude of the central correlation increases steadily with averaging time as the data-averaging process filters small-scale and baroclinic features (noise), so that monthly maps have the strongest correlations. They also discovered that for all regions of the United States, warmest conditions occur when a negative height anomaly is a few thousand km west to northwest of the region and positive height anomalies are within 1000 km of being overhead, usually to the east of the station. This means that the city either is under the heart of an anomalous ridge or is receiving strong southwesterly flow from the southwest quadrant of the ridge. Reversing the signs of the anomaly centers will produce the coolest conditions. In summer, local positive correlations between 700-mb heights and surface temperature are extremely important. All but five of the 109 stations have maximum positive correlation centers within 600 km of the city, and nearly 60% are within 300 km. In contrast, during winter, distances from station to correlation center are larger, ranging from 1000 to 2000 km. In summary, large-scale advection strongly affects surface temperature during winter, while local processes and smaller spatial scales dominate summer temperature.

Precipitation can also be specified, although results are always worse than those for temperature. Klein and Bloom (1987) specified United States winter and summer precipitation from a 79-point grid over North America and adjacent waters. For the monthly precipitation data, they used the 60 non-contiguous climatic divisions created by Englehart and Douglas (1985); these divisions are evenly distributed across the 48 contiguous states and are long-term averages of approximately 15 stations per division. Klein and Bloom found that, averaged over the United States, specification explains about 15% more precipitation variance in winter than in summer; that is, 45% versus 30%, respectively. This is consistent with the less coherent and more convective nature of summer precipitation. Pooling the climatic divisional data into the nine precipitation regions of Karl and Koscielny

(1982) explains about 12% more variance than the individual divisions within each region: winter rises to 57% and summer to 43% explained variance. This suggests that spatial smoothing minimizes small-scale and random effects, thus clarifying the relationships between precipitation and the large-scale circulation. Harnack and Lanzante (1985) specified United States area-averaged precipitation from 700-mb heights, North Pacific SSTs, and North Atlantic SSTs. Their results were strikingly inferior to those of Klein and Bloom. Nevertheless, they did manage to conclude that 700-mb heights are generally superior to SSTs for predicting precipitation.

Specification falls somewhere in the middle between the two approaches to synoptic climatology (Figure 1.1). In some ways, it is an environment-to-circulation approach to synoptic climatology because the selection of predictors (the circulation data) is not independent of the criterion variable (air temperature). However, because specification does generate a daily prediction of a surface environmental variable, model comparison statistics (Chapter 6) can be calculated, therefore suggesting that it is a circulation-to-environment approach.

Specification continues to evolve today. Klein and Walsh (1983) compared Klein's procedure, which uses forward stepwise screening to specify surface temperature from 700-mb height anomalies, with the specification technique of Walsh et al. (1982), which regresses the temperature anomalies onto the coefficients of the first few eigenvectors. They determined that, on average, Klein's procedure explains about 15% more of the winter-temperature variance than does this eigenvector-based strategy.

A more robust form of eigenvector-based specification has recently emerged which produces results at least as good as Klein's procedure. Hewitson and Crane (1992a) based their work on two hypotheses and one assumption. The hypotheses were: (1) GCMs accurately simulate present synoptic-scale circulation; and (2) eigenvector-based specification can estimate high-resolution surface temperature from these circulation characteristics. Their assumption was: with a doubling of CO_2, the underlying circulation modes will remain the same, but the frequency, persistence, and intensity of synoptic features will change. Growing from these three ideas, therefore, their goal was to specify surface temperature from GCM projections of a doubled-CO_2 world.

To test the first hypothesis, Hewitson and Crane used observed surface pressure and a present-day GCM simulation of surface pressure. They re-interpolated the GCM data to the same grid as the observed data and applied a running 13-day filter to the gridded data to remove variance longer than the synoptic time scale. Then they ran an S-mode, orthogonally rotated PCA on the correlation matrices. They did not cluster the results; this is not a map-pattern classification. Hewitson and Crane plotted and compared statistically the resulting components-loading patterns for the seven components retained in each PCA. The maps were nearly identical, suggesting that their first hypothesis is correct: the model simulates contemporary synoptic-scale circulation accurately.

They developed a stepwise polynomial-regression equation for each grid point in the observed data set to evaluate the second hypothesis: that surface temperature can be specified from the model output. The dependent variables are the scores on the seven circulation components for day t, $t-1$, and $t-2$, plus their other combinations. This produces 252 dependent variables to enter the stepwise procedure. Unlike Klein (1983 and 1985a), Hewitson and Crane did not try to minimize the number of variables entered into each equation; instead, they sought maximum explained variance by keeping all variables with contributions significant at the 99% level. The resulting equations for each grid point contain between seven and 26 variables. They applied the polynomial equations to component-score sequences generated by test data sets, and compared observed and predicted temperature. Their results show, on monthly time scales, that errors are roughly $\pm$ 1°C for most of the continental United States. Thus, the eigenvector-based specification can produce accurate temperature forecasts.

Hewitson (1992) used this strategy to focus on the relationships between daily circulation and grid-point temperature. He demonstrated that the history of the synoptic features (their sequencing) exert considerable control on surface temperature. He also applied the transfer functions to circulation output from a present-day GCM to see if model-specified temperature matched output from the model's normal temperature algorithm.

Hewitson and Crane (1992b) coupled this specification procedure with their assumption that synoptic patterns would be the same in a CO_2-doubled world to predict United States surface temperature. If this assumption is correct and the GCM circulation output is accurate, then their results show that the Great Plains–Rocky Mountains region will cool in winter, and that the West Coast will cool altogether with global warming. These results differ considerably from the surface temperature predicted by the GCM's regular temperature-determination scheme. Given the inherent problems with GCMs, the results of Hewitson and Crane should not be taken as an accurate portrayal of future climate. However, their methodology is sound and their specification procedure does appear to be a significant improvement over the eigenvector-based strategy of Walsh *et al.* (1982). Perhaps a head-to-head comparison with Klein's procedure is in order.

6 Relationships between the atmospheric circulation and the surface environment

The working definition of synoptic climatology used in this book demands that the investigator relates the atmospheric circulation to the environment. Until now, the emphasis has been on the atmospheric circulation and, more specifically, on classification of the circulation. Worked examples for the key synoptic-classification techniques have been presented. Now the weight changes to the environment and its association with the atmospheric circulation.

An important theme of the book is synoptic climatology's relevance to pressing environmental issues. In this chapter, to illustrate the value of synoptic climatology in environmental analysis, I apply the worked examples to several important environmental problems. To represent these problems, I develop four environmental scenarios (with five variables): urban-air quality (surface-ozone concentration); acid rain (wet-sulfate concentration); agriculture (corn yield); and fluvial hydrology (stream flow and in-stream sulfate concentration).

In the scenarios, each individual variable is meant to represent the more general state of the the environment or environmental problem. For example, photochemical processes in urban areas produce a suite of oxidants and other chemicals collectively termed smog. Surface ozone is the most abundant of these reaction products and is used as an index of the severity of photochemical air pollution (Comrie, 1990). In the urban air-quality scenario, I therefore treat surface-ozone concentrations as symptomatic of the state of the air quality in the Pittsburgh metropolitan area. As part of this holistic view, I introduce each scenario by presenting a summary of the environmental problem to be addressed by synoptic climatology.

This chapter has two related goals: (1) to determine how well each synoptic-classification technique relates to the surface environment; and (2) to ascertain which specific synoptic classification does the best job relating to the surface environment. To evaluate and compare the ability of the synoptic classifications to relate to the environment, it is important to have appropriate quantitative tools. Numerous measures have been developed to evaluate model effectiveness. In this chapter, I use a suite of model-performance statistics to rate the track records of the synoptic classifications with each environmental scenario.

The promise of synoptic climatology to environmental analysis can only be realized if the atmospheric circulation can be related to the surface environment and, more important, if that relationship helps the environmental scientist better understand how the environment works. A comprehensive synoptic climatology of each of the environmental scenarios is beyond the scope of this chapter. Nevertheless, to illustrate the power of synoptic climatology in environmental analysis, I find it worthwhile to discuss notable relationships between the atmospheric circulation and the environmental scenario that are revealed by the analyses.

Consequently, I divide this chapter into three sections. In *Model-performance statistics,* I present a suite of methods to evaluate and compare the circulation-to-environment classifications presented in Chapters 2–5. Next, I apply those statistics to the classifications and discuss the results in the section on *Environmental Scenarios.* I also present a scaled-down synoptic climatology of each environmental scenario. In the *Recapitulation* at the end of the chapter, I summarize these statistics and discuss the results of the classification evaluations and comparisons.

Model-performance statistics

A key goal of this chapter is to evaluate how well the worked synoptic classifications relate to the environmental scenarios. A corollary objective is to compare the performance of each classification to determine which one does the best job relating to the environmental scenarios. In a related series of articles, C.J. Willmott appraised several measures used to evaluate and compare geophysical models (Willmott, 1981, 1982, 1984; Willmott *et al.*, 1985). His primary objective was to establish "the degree to which model-predicted values approach a linear function of the reliable observations" (Willmott *et al.*, 1985; p. 8995).

Although Pearson's Product-Moment Correlation Coefficient (r), the coefficient of determination (R^2), and tests of their significance are commonly used in the literature for determining model performance, Willmott demonstrated that they are inadequate for this purpose. To satisfy this function, he recommended an alternative set of indices for comparing predicted and observed values, including the mean absolute error (MAE), root mean square error (RMSE) and Willmott's index of agreement (d).

MAE and RMSE estimate the average error and take the form:

$$\text{MAE} = N^{-1} \sum_{i=1}^{N} | P_i - O_i | \qquad (6.1)$$

and
$$\text{RMSE} = [N^{-1} \sum_{i=1}^{N} (P_i - O_i)^2]^{0.5} \qquad (6.2)$$

O and P are the observed and model-predicted variables, respectively, and N is the number of cases. Willmott describes MAE as the "actual average error," while RMSE can be considered a high estimate of MAE. In most instances, MAE and RMSE produce similar results, except that extreme values are inflated in RMSE because of the squaring of the ($P_i - O_i$) term. Nevertheless, RMSE is more widely used, perhaps because it is more amenable to mathematical decomposition and analysis.

Willmott demonstrated that RMSE can be partitioned into systematic (RMSE_s) and unsystematic (RMSE_u) terms:

$$\text{RMSE}_s = [N^{-1} \sum_{i=1}^{N} (\hat{P}_i - O_i)^2]^{0.5} \tag{6.3}$$

and

$$\text{RMSE}_u = [N^{-1} \sum_{i=1}^{N} (P_i - \hat{P}_i)^2]^{0.5} \tag{6.4}$$

$\hat{P}_i = a + b\, O_i$ where a and b are the parameters associated with an ordinary least-squares linear regression between O and P. RMSE_s is the model-induced error; RMSE_u is error that cannot be accounted for by the model. Because a good model should explain most of the systematic variation in the observed variables, RMSE_s should be small, while RMSE_u should approach RMSE. If RMSE_s is relatively large, the model can be tuned to decrease this model-induced error. The percentage of the error that can be eliminated by fine tuning is:

$$(\text{RMSE}_s / \text{RMSE})^2 \tag{6.5}$$

RMSE_s and RMSE_u are good to use if goals of the performance evaluation include: (1) selecting the best model among a suite of models; and (2) deciding whether or not a model is likely to be improved by tuning.

Based on knowledge of the error (RMSE), Willmott developed a measure that reflects the relative degree to which predicted values approach observed values.

$$d = 1 - \frac{N\,(\text{RMSE})^2}{\text{PE}} \tag{6.6}$$

where the potential error variance is:

$$\text{PE} = \sum_{i=1}^{N} [|\,P_i| + |\,O_i\,|]^2 \tag{6.7}$$

Willmott's *index of agreement* (d) varies between 0.0 and 1.0, with 1.0 expressing perfect agreement between observed and predicted values and 0.0 describing complete disagreement. The index of agreement is a measure of the degree to which a model's predictions are error free. There is no absolute value that the magnitude of d must reach to become "significant". Instead, the investigator must evaluate the import of the d-score on the basis of his or her knowledge about the phenomenon being studied, the data's accuracy, and the model being employed. The index of agreement is only meaningful in the context of the problem under investigation.

To calculate the model-performance statistics, the investigators computed the mean of each variable (for example, surface-ozone concentration) associated with each synoptic-type or map-pattern class. These means became the *predicted* value for each day having that synoptic class; the real data formed the *observed* time series. The predicted time series was explicitly not intended as a hindcast, but did show which classification scheme provided the best separation and sequencing of the variable values, thereby producing superior model-performance scores.

I present RMSE, $RMSE_s$, $RMSE_u$, $(RMSE_s/RMSE)^2$ and d for the five worked examples that used the circulation-to-environment approach to synoptic climatology. Model-comparison statistics cannot be calculated for environment-to-circulation classifications such as the worked example of compositing. For each environmental variable, I use these statistics to rate the performance of the circulation-to-environment synoptic classifications. The investigative team calculated MAE in all cases, but, because it did not provide additional information, I only show RMSE.

The environmental scenarios

Urban air quality

Air-quality problems beset many of the world's cities. One of the most severe pollutants is ozone, which can affect human health, vegetation, and various materials, not only in the urban environment but downwind in rural areas. Ozone (O_3) is the triatomic form of the more familiar diatomic oxygen molecule (O_2). Ozone is an oxidant because its extra oxygen atom readily reacts (oxidizes) with inorganic and, especially, organic substances. Surface ozone occurs naturally, but potentially dangerous concentrations are the result of human activities. In the following brief summary of the surface-ozone problem, I rely heavily on the review of Comrie (1990) in which he explains ozone behavior, sources, and transport. See his references for the literature on the subject.

Ozone is a secondary pollutant. It is not emitted directly to the environment but, instead, is formed by complex reactions among the primary pollutants, known as precursors. The precursors of ozone include various nitrogen oxides (NO_x) and volatile organic compounds, mainly hydrocarbons. In the presence of sunlight, the precursors react photochemically to form ozone and other chemicals. Because ozone formation is dependent on photochemical activity, ozone concentrations are linked intimately to meteorological conditions. For example, maximum ozone formation occurs on sunny summer days.

The atmosphere produces surface ozone naturally. Much of the natural (background) ozone is transported from the stratosphere to the planetary boundary layer. These *tropopause-folding* events occur routinely with frontal and convective processes. None the less, human production of ozone's precursors is necessary for surface-ozone concentrations to reach harmful levels. Many kinds of industrial facilities, such as petrochemical plants and fossil-fuel power stations, produce large volumes of volatile organic compounds, while the chief generator of NO_x is motor vehicles. Thus, the high densities of industry and vehicles associated with urban areas make cities the main source of the precursors and, ultimately, of ozone.

Ozone is not restricted to the cities where it is generated. Ozone plumes carry the oxidant many tens of kilometers downwind of cities, with highest ozone concentrations occurring not in the cities but in their lee. Long-range transport is also a concern. Over the course of days, precursors can move out of a source region, evolve into ozone, and concentrate hundreds to thousands of kilometers away. Worse, multiple sources can focus their pollutants into one area. In the worst case, a region where ozone concentrates can have significant urban concentrations of its own, compounding the problem. For instance, ozone-caused forest damage in western Pennsylvania relates to precursors generated in the southern United States and Ohio River Valley, as well as to ozone and its precursors coming from Pittsburgh (Comrie, 1992a).

The temporal and spatial scale of medium- to long-range transport is, therefore, the synoptic scale. Investigators have directed considerable effort to understanding ozone formation associated with individual synoptic-scale high-ozone events but have conducted only a few studies that approach the longer time-scales typical of most synoptic climatologies. Yet, the holistic, synoptic-climatological approach has much to offer surface-ozone research.

Encouraged by this review of the literature, the investigative team chose to study the relationships between synoptic-scale atmospheric circulation and ozone concentrations in metropolitan Pittsburgh. As noted in the introduction to this chapter, ozone concentrations in the urban atmospheric environment are an index of a complex suite of photochemical-reaction products. Therefore, the synoptic climatology of ozone should reveal large-scale atmospheric controls on urban air quality.

The investigators obtained hourly ozone data from the National Air Data Branch of the United States Environmental Protection Agency (EPA). No single site in Pittsburgh has sufficient data to cover the 10-year study period, 1978–1987. However, four sites have extensive daily records that, if pooled, do cover this span. Correlations of the four sites' ozone data show that they appear to be part of the same regional ozone regime, paralleling each other under a range of synoptic conditions. Consequently, the investigators pooled the data from the four sites to provide a single regional ozone time series, somewhat free from local noise. The pooled data characterize ozone over the metropolitan region, with sites located in urban, suburban, and park settings of the city's northern and eastern sectors.

The data are daily maximum-hourly ozone concentrations. The investigators settled on the maximum, rather than the mean, for two reasons. First, it is a more sensitive indicator of limiting environmental and meteorological conditions than the mean. Second, the maximum is the variable utilized in regulatory standards.

The daily data were standardized using the conventional Z-transformation, then pooled to form an average Z-score for the four sites. The data were then "destandardized" by reversing the Z-transformation calculation. This created a single time series of pooled daily maximum-hourly ozone concentrations for the metropolitan area. Missing data from one or more sites introduces some non-systematic bias into the time series. The final time series includes 3089 days for analysis; that is, 15% of the data are missing. Only one site represents the region in some cases, and there are two significant wintertime gaps when none of the sites have data. See Comrie and Yarnal (1992) for more details of the data and their treatment.

Synoptic-classification performance. In the following, first, I rate the *overall performance* of the circulation-to-environment approaches for the urban air-quality scenario and, second, I compare the *relative performance* of the five individual circulation-to-environment classifications. I follow this same procedure in the other three environmental scenarios.

When taken together, the model-performance statistics show that all of the classifications did a very good job explaining the variations in Pittsburgh's ozone (Table 6.1). For all classifications, the average error estimate, RMSE, is moderate. $RMSE_u$ makes up a large proportion of RMSE, thus suggesting that the classifications are able to predict without excessive error. The ratio $(RMSE_s/RMSE)^2$ indicates that about one-third of the error can be eliminated by fine tuning the classifications. The values of the index of agreement, d, are relatively high for synoptic-climatological classifications.

There is no clear winner, however, when the five classification schemes are compared. The manual classification produced the highest d-value, lowest overall error, and least systematic error. In addition, this classification is moderately tunable, implying that even more error could be removed with improvements to the classification scheme. Nevertheless, the next best performer, the correlation-based map patterns, was only marginally inferior (that is, only one- or two-hundredths shy) on all measures. The third-place finisher, the circulation index, was just a couple of hundredths short of the winner in all measures. With these scores, an investigator could choose the classification he or she feels most comfortable with; the results of the synoptic climatology would be comparable to those produced by the other schemes.

Relationships between the atmospheric circulation and the urban air-quality scenario. Because the manual synoptic types produced slightly better marks in the model-performance tests, they were used to study the synoptic climatology of surface-ozone concentrations in metropolitan Pittsburgh. In the following, I concentrate on four aspects of the atmosphere–ozone relationship: average, extreme-event, between-season, and year-to-year conditions. These findings are distilled from Comrie and Yarnal (1992). Please refer to Chapter 2's worked example for background on the synoptic types.

The nine synoptic types of the manual classification each possess characteristic mean levels of ozone. Each synoptic type also has a different frequency of occurrence. Multiplication of the means by the frequencies

Table 6.1 Synoptic-classification model performance in the urban air-quality scenario

	Manual synoptic types	Correlation-based map patterns	Eigenvector-based synoptic types	Eigenvector-based map patterns	Circulation index
RMSE	1.20	1.22	1.25	1.24	1.22
$RMSE_u$	0.96	0.96	0.98	0.97	0.97
$RMSE_s$	0.72	0.74	0.78	0.77	0.74
$(RMSE_s/RMSE)^2$	0.36	0.37	0.39	0.30	0.37
d	0.57	0.56	0.52	0.52	0.55

produces the *cumulative dose* of ozone maxima. The departure from the mean cumulative dose is an indicator of a synoptic type's *relative* positive or negative influence on the ozone climatology.

The most important low-ozone type is RC. Weather conditions on RC days are generally cool and cloudy to partly cloudy, with some precipitation. These conditions are unfavorable for ozone formation, thus explaining the low-to-moderate mean maximum-ozone values. Nevertheless, RC produces 22% of the total ozone dose in Pittsburgh, largely because of its high frequency (almost 90 times per year), rather than its moderately-low mean maximum-ozone concentration.

The highest maximum-ozone concentrations of 55 ppb and 51 ppb, respectively, are experienced on typical BH and EH days. Both types occur roughly 60 times per year. As a result, BH produces almost 25% of the total cumulative dose, while EH is responsible for about 20%. Under both types, the warm, sunny conditions and low wind speeds, with southerly to westerly long-range transport over numerous precursor sources, are highly favorable for ozone production in summer.

Over the 10-year study period, there were 30 days on which maximum-ozone concentrations exceeded the EPA safe-ozone standard of 120 ppb. Not surprisingly, 22 of these events occurred on EH and BH days. EH is associated with nearly twice as many high-ozone events as BH, suggesting that dangerous ozone concentrations in Pittsburgh are more likely to arise under the weak pressure gradients and low wind speeds of EH than the relatively more disturbed conditions of BH. Five high-ozone events occurred on CF days, reflecting this type's kinship to BH and the influence of the potentially ozone-rich warm sector preceding the front. A composite high-ozone event (Figure 5.1) strongly suggests a combination of the pressure patterns representing the BH and EH pressure configurations.

For contrast, 30 days below 5 ppb were deemed extreme low-ozone events. The majority of these occurred under RC conditions. CF, which was associated with five high-ozone events, is also associated with five low-ozone events. The moderate mean-ozone value of CF reflects this dual nature. Evidently, the combination of warm- and cold-sector air masses in this type creates a statistical entity that does not reflect accurately the discontinuity or steep gradient in ozone concentrations that exists in the frontal zone. Thus, the maximum ozone value recorded in Pittsburgh on CF days is a function of a number of factors, most notably frontal position and pressure and ozone gradients.

Monthly mean maximum-ozone concentrations show a distinct annual cycle, with values over 70 ppb in July and about 15 ppb during December and January. Similarly, the synoptic types display strong seasonality in terms of frequencies, mean maximum-ozone concentrations, and cumulative doses. Of the categories most important to high-ozone events, BH is principally a summer type. EH and CF, also important to high-ozone values, normally occur in all months but peak in summer through autumn. The most frequently occurring low-ozone type, RC, is common throughout the year, but dominates winter and spring, when ozone levels are lowest.

There is considerable interannual variability of synoptic-type frequencies, mean maximum-ozone concentrations, and cumulative doses of ozone. One important aspect of interannual climatic variability is the annual changes in the frequencies of different weather systems; these in turn contribute to variations in mean annual-ozone concentrations. However, interannual variations in mean ozone levels within each synoptic type (within-type variations) can potentially amplify or counteract the effect of frequency changes. Nevertheless, for the period 1978–1987 in Pittsburgh, interannual variations in the mean ozone concentrations for most synoptic types mimic the overall mean ozone concentrations.

The cumulative doses of each type reflect the combined effects of varying frequencies and means over time. The relative importance of EH and BH as high-ozone types alternates several times over the study period, in roughly opposite phase. A similar opposite-phase relationship exists between the low-ozone types, RC and PH$_w$. EH is the dominant high-ozone type and RC the dominant low-ozone type in high-ozone years; conversely, in years with lower mean maximum-ozone concentrations, BH dominates the high end and PH$_w$ the low end. This makes physical sense: BH–PH$_w$ years tend to have more fast-moving, zonally migrating storm systems; EH–RC years possess slower-moving, meridionally migrating systems or standing long waves. Thus, lower ozone levels are associated with the more disturbed conditions observed with strong zonal flow, while higher levels result from the more stagnant conditions accompanying meridional flow.

However, interannual differences in the synoptic-climatological control of ozone levels may not be important to annual variations in Pittsburgh ozone. The effects of interannual variations in synoptic-type frequencies can be removed from the mean maximum-ozone data by a relatively simple matrix manipulation (Comrie, 1992c). The resulting time series suggests that different annual frequencies of weather systems over Pittsburgh have surprisingly little influence on year-to-year changes in mean maximum-ozone concentrations. It also appears that the overall influence of zonality or meridionality is relatively small. Although the evidence is circumstantial and only covers a 10-year period, much of the interannual variation in ozone appears to be attributable to non-climatic factors. Emissions changes are the most likely cause.

This somewhat conventional synoptic climatology can be fine tuned by *sequencing*. The composite Figure 5.1 illustrates the explanatory power of a sequence of days. However, a drawback of compositing is that, because it uses an environment-to-circulation approach to synoptic climatology, it only works when there is a clear association between the climate and the environmental variable. Unless the investigator hits upon such a relationship by luck, he or she must be aware of the association and select specific criteria that will draw out the link between the circulation and environment. If maps for low or moderate ozone are composited, for example, ambiguous maps result because of the variety of synoptic types associated with those ozone levels. To get around this problem and take advantage of the increased explanation afforded by synoptic-type sequences, Comrie (1992b) devised a simple technique to identify all possible synoptic-type sequences in any circulation-to-environment scheme.

Using the Pittsburgh air-quality scenario, he showed that this technique includes information on climatically significant sequences of synoptic categories, the persistence or speed of systems, the origins of air masses, and more. Thus, sequencing dramatically improves the utility of circulation-to-environment approaches to synoptic climatology.

In summary, the manual synoptic-typing scheme produced a convincing synoptic climatology of surface-ozone concentrations. Still, based on the model-comparison statistics, it is likely that any of the other circulation-to-environment approaches also would have produced valid results. Additionally, fine tuning the classification, which in this case involved sequencing the synoptic types, improved the results of the synoptic climatology.

Acid rain

Yarnal (1991) provides a relatively comprehensive survey of climatological research on acid rain. He shows that the synoptic-climatological perspective provides insights into the relationship between pollution sources and acid rain not afforded by other approaches. The following is a slightly modified version of the introduction to that paper. For clarity, I do not provide references here; see Yarnal (1991) for the literature on the subject.

Acid precipitation, more commonly called acid rain, is the direct result of a complex chain of events that takes place in the atmosphere. The sequence starts with the emission of primary pollutants and the more important precursors of secondary pollutants such as the oxides of sulfur and nitrogen. Transport and diffusion of these materials follow. During transport, acids or acidifying ions result from transformation of the precursors. Precipitation eventually scours both primary and secondary pollutants from the atmosphere and deposits them at the surface as acid rain.

Atmospheric processes control transport, transformation and precipitation: the meteorological condition determines the direction and dispersal of the materials after emission; the amount of available sunlight, heat and moisture regulates the types and quantities of secondary pollutants produced; and the type of weather determines whether precipitation will occur and the amount of pollution cleansed from the air. Thus, the production of acid rain varies with the state of the atmosphere.

Research on acid rain by atmospheric scientists can be broken into two streams, meteorological and climatological. Meteorological studies center on the instantaneous or short-term physics and chemistry involved in producing acid rain. The meteorological approach is theoretical and process-oriented; the current emphasis is on mathematical modeling. Much of this work has focused on smaller-scale processes. In contrast, climatologists study the long-term, large-scale relationships between the atmosphere and acid rain. The primary concern of climatological research is to explain the variations and trends in acid rain over time and space. This strategy is empirical, based on the statistical analysis of observed phenomena.

Research on the climatology of acid rain has both strengths and weaknesses. The main strengths relate to three perspectives of the climatological approach: that is, the statistical, spatial and temporal viewpoints. Statistical averages isolate and clarify atmospheric processes that appear to be random when viewed as individual meteorological events. In addition, statistics on both the mean state and variability of the acid rain–atmosphere connection provide a foundation for studies on the theory and environmental effects of acid rain.

Much climatological research focuses on the spatial dimensions of atmosphere–environment interactions. In climatological analysis, the expression of some environmental processes varies with spatial scale. For acid rain, secondary pollutants are most evident at the synoptic scale. This implies that the synoptic-climatological approach is appropriate for studying the association between the atmosphere and secondary pollutants.

Perhaps the best reason for investigating the climatology of acid rain relates to temporal variability. Holding emissions constant, a change in precipitation amount alone will produce changes in acid-rain deposition levels. Also, even if precipitation totals did not fluctuate annually, interannual variations in the types and timing of storm events would still have a significant impact on the average annual acidity of precipitation. Thus, amounts and concentrations of acid rain will vary from year to year because of climatic variability.

In areas near pollution sources, separating natural variation due to climate from variation attributable to emissions changes is a major problem in evaluating trends in acid rain. In contrast, in areas where long-distance transport is the major source of pollutants (that is, areas remote from sources), climate is believed to be largely responsible for variations in acid rain.

Despite the strong points of climatological studies, there are several weaknesses that plague them:

— Insufficient data in both time and space;
— A lack of concurrent measurements of significant parameters that define the source-receptor relationship (for example, data may be daily, weekly, and event-based);
— Spatial and temporal representativeness of the data;
— Measurement uncertainty;
— Physical and chemical reasonableness of the derived relationships.

The first four of these problems relate to data. Because climatological assessments use observations, the strength and clarity of the climate–acid rain relationship depend on the quality, quantity and timing of the data. The working assumption is that the data are good; the reality is, because of one or more of the above difficulties, they are often poor. Much uncertainty over the atmospheric control on acid rain stems from these data problems.

The fifth point is true of all climatological analyses: significant statistical associations between variables do not prove causation. Still, such relationships do suggest possible physical or chemical mechanisms and, therefore, are invaluable for providing the basis of subsequent theoretical research.

With this background, the investigative team felt assured that variations in the atmospheric circulation would relate well to acid-rain variables. To serve as a surrogate for the total complex of physical and chemical processes associated with acid rain, they obtained sulfate-concentration data from the Penn State Scotia station of the MAP3S precipitation-chemistry network (The MAP3S/RAINE Research Community, 1982). Precipitation samples are measured on an event basis, with an event defined as any 24-hour period during which precipitation has occurred.

A unique attribute of the Penn State site is the rural setting in central Pennsylvania. All significant point sources of pollution are more than 40 km from the sampling site. The major pollution sources in the prevailing upwind direction are more than 100 km away. Thus, the data are regionally representative. An interesting note is that the Penn State site has consistently registered the second-most-acid precipitation in the national network in recent years. The most acidic readings come from a station also run by Penn State in an adjacent county.

Synoptic-classification performance. The acid-rain scenario (Table 6.2) presents results that are quite different to those for the urban air-quality scenario. Recall that the circulation-to-environment classifications all did a good job in relating the circulation to surface ozone; that is not the case in the acid-rain example. Here, RMSE is generally lower than in the air-quality scenario, denoting less average error. Unfortunately, $RMSE_u$ is quite low relative to $RMSE_s$, suggesting that much of the error is classification-induced. With the very high percentage of error that could be eliminated by fine tuning, as shown by the values of $(RMSE_s/RMSE)^2$, the implication is that the investigators should devote considerable effort to tuning these generic classifications to the specific needs of acid rain. The d-scores appear to be moderately low in relation to those observed in the air-quality scenario. Because the sulfate concentrations are precipitation-borne, and precipitation is notoriously difficult to model, these values are not bad and should not be used to dismiss the synoptic classifications. In total, I find these scores encouraging. Acid-rain scenarios should be addressed by circulation-to-environment approaches to synoptic climatology, although special effort must be made to refine the techniques to match the temporal and spatial scales of the phenomenon.

Two classification schemes clearly outdistance the others in the acid-rain scenario. The eigenvector-based map patterns and manual synoptic types had d-scores roughly 100% and 60% higher, respectively, than the other three classifications. Additionally, these two had slightly lower average error values. None the less, the high squared ratio of $RMSE_s$ to RMSE suggests that the improvements needed in these two classifications is comparable to those of the others. Furthermore, it is uncertain whether the relative distance between d-scores and rankings would be maintained after fine tuning. At this point, however, it seems safer to choose either of the top two classifications over the others. The six-point spread between the

Table 6.2 Synoptic-classification model performance in the acid-rain scenario

	Manual synoptic types	Correlation-based map patterns	Eigenvector-based synoptic types	Eigenvector-based map patterns	Circulation index
RMSE	0.98	1.00	1.00	0.96	1.00
$RMSE_u$	0.25	0.15	0.18	0.28	0.16
$RMSE_s$	0.94	0.99	0.98	0.92	0.98
$(RMSE_s/RMSE)^2$	0.96	0.90	0.96	0.92	0.96
d	0.31	0.18	0.20	0.37	0.19

eigenvector-based map patterns and the manual synoptic types gives the nod to the eigenvector-based classification scheme.

Can the differences among the classification performances be explained? The low scores of the eigenvector-based synoptic types and circulation index are understandable. The investigators based the former's PCA on a suite of variables that did not include precipitation and, consequently, the classification was not attuned to it. The circulation index is based on pressure-pattern intensity. A high acid-rain event often calls for a sequence of low-gradient, high-pressure days, followed by high gradients and low pressure. This sequence regularly takes about three days to unfold; if it did get compressed into one day, the value of the synoptic index would be difficult to predict. The relative failure of the correlation-based map patterns is surprising, especially given the success of the other map-pattern-based schemes. Although I can speculate on the causes of this flop, these guesses would be difficult to prove without experimentation. One unsettling possibility is that all map-pattern classifications are unstable and that an investigator can never conduct this type of synoptic climatology feeling confident that the classification will perform well.

Relationships between the atmospheric circulation and the acid-rain scenario. The model-comparison statistics show that the eigenvector-based map patterns (Figure 4.7) do the best job relating the atmospheric circulation to the Penn State precipitation-sulfate concentrations. Accordingly, a synoptic climatology of the acid-rain scenario is being prepared using this classification scheme. Although the research is not complete, it is possible to make a number of generalizations about high- and low-sulfate events and about interannual variations based on the eigenvector-based map-pattern classification.

Of all days with elevated precipitation-sulfate totals, nearly half exhibit some form of Bermuda high. Map Pattern 1, a strong Bermuda high, occurs on one-third of all high-sulfate days, and Map Pattern 6, a much weaker Bermuda high, is present on about 15% of such days. Only four more of the remaining 10 map-pattern classes are important on high-sulfate days. Map Pattern 2, which displays a low-pressure center over the Ohio River Valley and is critical to low-sulfate events (discussed below), is observed on 15% of the high-sulfate days. Map Pattern 4 also appears on 15% of the high-sulfate days. It manifests a high-pressure pattern over the midwest, and implies that the frontal system which has just passed through the study area will be replaced by high pressure. Map Patterns 3 and 5 are each associated with 10% of the high-sulfate events; the former has high pressure centered over the Great Lakes, while the latter is a Great Lakes low. In summary, with the exception of Map Pattern 3, all of these configurations obviously relate to precipitation. Acid rain, associated either with the conditionally unstable, humid air in the Bermuda high's northwest quadrant or an organized low-pressure system, by definition must fall on the event day.

The map patterns preceding a precipitation event determine its acidity. Two days before a high-sulfate event, the three most frequent map patterns are 1, 3 and 4. Map Pattern 2, which washes pollutants out of the air, almost never occurs. The day before a high-sulfate event, different patterns dominate. Map Pattern 1 is relatively less frequent. Instead, high-pressure patterns 3, 6, 8 and 9 are more common than normal, occurring nearly 60% of the time. Taken together, these patterns demonstrate that high pressure and dry conditions are critical to sulfate build up in the atmosphere before an acid-rain event.

The above implies that high precipitation-sulfate levels are produced when key sequences of map patterns create optimum conditions for sulfate formation and transport. On the one hand, single-day circulation-to-environment classifications do not clearly portray such a sequence. Sequencing techniques would probably alleviate this problem (Comrie, 1992b). On the other hand, environment-to-circulation approaches, such as compositing, do an excellent job picturing multi-day processes.

Composite maps of surface pressure associated with in-stream high-sulfate events (Figure 6.3; covered later in this chapter) are similar to those for acid-rain events. However, the day of the precipitation event often precedes the day with highest in-stream sulfate values. Thus, the day before the stream's high-sulfate event is the day of the acid-rain event; two days before the stream's high-sulfate event is the day before the acid-rain event; and so forth. The composite sequence for high-ozone events in Pittsburgh (Figure 5.1) is also similar to the succession of map patterns associated with acid rain.

Two to three days before a high precipitation-sulfate event, strong high pressure sits over the East Coast. On many days, this will be the Bermuda high (Map Patterns 1 and 6), but on others it will be an elongated ridge (Map Pattern 8), Ohio River Valley high (Map Pattern 9), or Great Lakes high (Map Pattern 3). Pollutants generated in the Ohio River Valley and other regions will stay near their source or lazily drift toward west-central Pennsylvania. The strong solar radiation, plus relatively high temperatures and humidity, favor the transformation of sulfur and nitrogen oxides into acids or acidifying agents. On the day before an event, the high starts to move offshore, a low approaches from the west, and pressure gradients increase between these two centers. This stimulates strong south to southwesterly airflow, while SO_x and NO_x transformation continues. Thus, primary and secondary pollutants are funneling toward Penn State. On the event day, the high often moves completely offshore and a storm system enters the area, washing the pollutants out of the atmosphere as acid rain. Alternatively, the high moves more slowly, but, because of increasing upper-level support and continued surface heating, the conditionally unstable air overturns and localized thunderstorms drop acidified rain on the study site. The former is the normal sequence during the cooler months of the year, while the latter is typical of the summer months. In summary, the combination of circulation-to-environment and environment-to-circulation classifications complement one another. This fusion greatly enhances the understanding of patterns and processes produced by either one of the synoptic-classification approaches.

Low precipitation-sulfate events are much easier to explain with just the map-pattern classification. Low sulfate totals are associated with map patterns that produce precipitation, heavy cloud, and perhaps strong northerly wind one or two days before an event. Precipitation generated by the low-pressure systems of Map Patterns 2, 5, 10 or 12 clear the air, thus making possible the low-sulfate concentrations on the day or days to follow. This is especially true of precipitation events with strong northerly winds, such as those associated with lake-effect precipitation and cloud in the study area. For example, Map Pattern 7 and its powerful northerly flow is *never* associated with a high-sulfate event on ensuing days, but it is often a precursor to low-sulfate precipitation. On the day of the low-sulfate event, almost any map pattern can occur because the air has been cleared. Nevertheless, persistent map patterns that produce precipitation on days preceding the low-sulfate event are much more likely to occur.

Interannual variations in map-pattern frequencies can partially explain swings in precipitation-sulfate concentrations at Penn State. For example, 1981 had the highest sulfate concentrations during the study period, while 1983 had the lowest. Surprisingly, the difference did not occur in summer, when sulfate concentrations are highest in the northeastern United States and dominate the annual signal. Instead, concentrations were abnormally high in each of 1981's other three seasons, roughly doubling the concentrations of those seasons in 1983. The summer figures were somewhat typical in both years. Parallel to this, the Bermuda high of Map Pattern 1 was more frequent in 1981 than any other year, while it recorded its third lowest total in 1983. Map Pattern 2, the Ohio River Valley low, also showed less dramatic, but complementary fluctuations. These differences came primarily during the winter and spring. In general, Map Pattern 1 is relatively uncommon in April. In 1981, however, 13 days had this configuration, while not one Map Pattern 2-day occurred; these exact numbers were also posted in June, although the Bermuda high is much more common in that month. In the first half of 1981, Map Pattern 1 was witnessed 46 times (25% of all days) to just 12 observations of Map Pattern 2. In contrast, the first six months of 1983 saw Map Pattern 1 on only 19 days (10% of the time) and Map Pattern 2 on 28 days. Thus, it appears that increased (decreased) frequencies of the high-sulfate map pattern and concomitant decreases (increases) in the low-sulfate map pattern control acid rain at Penn State.

Unfortunately, it is not that simple. In autumn 1981, sulfate concentrations in precipitation exceeded those of summer, while in autumn 1983 they were about half the concentration of the preceding season. This cannot be explained by changes in map-pattern frequencies. In fact, Map Pattern 1 was much more common in autumn 1983, occurring 17 times in October through December, as opposed to just 12 times in autumn 1981. Similarly, Map Pattern 2 was observed 9 times in autumn 1981, and on only 8 days in autumn 1983. No other map pattern deviated significantly from its normal frequencies during these seasons. Thus, high sulfate concentrations in 1981 appear to result from a combination of pressure patterns favoring transport and transformation of the acid-rain precursors into the region but also higher ambient sulfate totals under all weather conditions. In contrast,

1983 had fewer map patterns favoring acid rain, but lower ambient atmospheric sulfate. These data suggest, therefore, that year-to-year fluctuations in precipitation acidity in the northeast have a climatic and an anthropogenic component. Using synoptic climatology, investigators can separate the climatic component from the anthropogenic signal (Comrie, 1992c) and more accurately assess society's contribution to the acid-rain problem.

Agriculture

The relationship between climate and agriculture is of fundamental importance to society. In particular, variation of crop yields with climate is particularly worrisome. Humanity has long realized that yields are poor when weather fails to cooperate. However, the exact nature of the association is unclear because of the complex relationships among the physical and cultural variables influencing agriculture.

None the less, there is a strong statistical relationship between climatic variation and corn (maize) yield. (Dilley, 1992, provided the information given below. For a more complete summary of the climate–corn yield literature, see that article.) In particular, multiple regression analysis shows that key climate variables explain a large proportion of the variance in corn yield. This research suggests that yields in the corn belt of the United States will be greatest under the following conditions:

— Average June temperature;
— Below-average July and August temperature;
— Average precipitation in the September-through-June period preceding the harvest;
— Below-average June rainfall;
— Above-average July rainfall.

The most important variable is July rainfall. In physiological terms, this suggests that corn yields are inversely related to moisture stress in high summer.

At least four problems, however, haunt multiple regression studies. First, the number of predictors (climate variables) may be greater than the number of corn-yield observations. Second, important predictors may be left out of the model. Third, the predictors may be highly correlated. Finally, although multiple regression is a linear statistical model, the relationship between climate and corn yield is likely to be non-linear.

In recent years, more sophisticated statistical techniques, such as PCA, eliminate some of these problems. For instance, PCA reduces the number of variables and eliminates the problem presented by correlations among variables. Unfortunately, advanced statistics have not overcome other problems such as missing predictors and non-linearities.

Synoptic climatology presents an alternative approach to climate–crop yield relationships. The holistic, synoptic approach represents the totality of climate, not individual variables, so that the non-inclusion of certain variables may be less important than in statistical models. Furthermore, the fuzziness inherent in synoptic climatology allows significant non-linearities to exist in the linkages between the large-scale atmospheric circulation and the micro-scale of the individual corn plant. Although many find such black box approaches unsatisfactory because there is no guarantee that the uncovered relationships are biologically meaningful, current modeling approaches to yield prediction are not significantly better in this respect.

The investigative team therefore determined that they would apply synoptic climatology to Pennsylvania corn yields and compare the results to those from the standard multiple regressions used by agricultural experts (see Dilley, 1992). They obtained corn yields from the *Crop and Livestock Annual Summary* published by the State of Pennsylvania. The *Summary* issues average per-acre yields of corn harvested in The Commonwealth. Six counties of the Southwest District and seven counties of the West Central District surround the Pittsburgh metropolitan area. The investigators carried out analyses at both county and district levels to look for differences in these two spatial scales.

Because the study period encompassed the 1978 to 1987 growing seasons, only 10 per-acre corn yields were available for analysis. The investigators compensated for the small temporal sample size through spatial replication. Each model comparison was run 15 times; 13 for the individual counties, plus two for the aggregated district-level yields.

The *Crop and Livestock Annual Summary* also publishes monthly temperature and precipitation averages and departures. Three National Weather Service stations in the Southwest Plateau region, which encompasses the 13 Southwestern and West Central counties, were employed in this study. Multiple regression using these climatological data with the yield data formed a benchmark for evaluating the effectiveness of the synoptic corn-yield predictions.

Synoptic-classification performance. The agriculture scenario took a divergent approach to the model assessment. First, instead of using the entire suite of synoptic classes to predict corn yields, the investigative team selected only the best predicting class. This was necessary because, with only 10 data points, there is a danger in having too many predictors. The investigators identified the best predictor by entering the synoptic-category frequencies for July, the climate of which is critical to corn yield, into a stepwise multiple regression. The regression chose the following categories (of the parenthetical classification scheme): RC (manual classification); Map Pattern 11 (correlation-based classification); Type 3 (eigenvector-based synoptic-type classification); Map Pattern 8 (eigenvector-based map-pattern classification); and circulation-index category 9. The investigators determined the circulation-index categories by dividing the range of index values (-0.961 to 0.685) into 14 class intervals, starting with the lowest index value. The ninth class approximates the mean pressure pattern for the standard-deviation interval $1.0 >$ Index > 0.5 (Figure 5.2), which looks like the mean pressure pattern associated with eigenvector-based Type 3. Second, the investigators made calculations for all 13 counties of southwestern Pennsylvania, plus the two aggregated districts, because there were only 10 years of crop yields. The data in Table 6.3 are the average values for the 13 counties; the district statistics were comparable (see Dilley, 1992, for more details).

The overall results are excellent. In the regression analyses, the first-chosen synoptic classes consistently explain far more variance than the traditional regressions based on weather variables (Dilley, 1992). Average error estimates are quite high, but the small sample size accounts for this. $RMSE_s$ is a relatively high proportion of RMSE, and $(RMSE_s/RMSE)^2$ scores show that the models could benefit from extensive tuning. Given the one-predictor strategy, however, this is not surprising. The d-scores are all impressively high, especially considering the small sample size.

Comparison of the classifications shows that one clearly outdistances the others: eigenvector-based synoptic typing. It has the lowest average error, requires the least tuning, and has an index of agreement nearly 13% better than its nearest competitor. The next two classifications are the map-pattern classifications. Although the numbers put the correlation-based classification ahead of the eigenvector-based scheme, the small sample size and pooling suggest that the difference may not be significant. Manual typing and the circulation index plainly lag the others; each of the key measures puts them at the bottom.

This three-tiered result can be rationalized. Dilley (1992) believes that the eigenvector-based synoptic-typing technique does an excellent job because plants respond physiologically to air-mass characteristics (for example, temperature, humidity, cloud cover, etc.), rather than to the wind direction and atmospheric pressure emphasized by map patterns. If this is true and map-pattern classification is not the best strategy for corn-yield prediction, the correlation-based and eigenvector-based map-pattern schemes still performed well. This suggests that the holistic design of map-pattern analysis appears appropriate to this task, perhaps because map patterns do give the investigator an idea of air-mass placement. It is easy to understand why the synoptic index did relatively poorly: pressure gradients and intensities do not necessarily delineate the air-mass characteristics that appear to be important to corn yield. In contrast, the manual classification's relatively inferior performance is more difficult to assess. The fact that, when classifying the daily weather maps, the investigators paid particular attention to precipitation and air-mass characteristics should make the manual scheme particularly well suited to this analysis. Still, the d-score of 0.61 is higher than than the highest value logged in any of the other scenarios.

Relationships between the atmospheric circulation and the agriculture scenario. As noted above, regression analysis shows that the basic relationships among temperature, precipitation, and corn yield reported in the literature hold true in southwestern Pennsylvania. July and August

Table 6.3 Synoptic-classification model performance in the agriculture scenario

	Manual synoptic type RC	Correlation-based map pattern 11	Eigenvector-based synoptic type 3	Eigenvector-based map pattern 8	Circulation index Class 9 (see text)
RMSE	9.32	8.26	7.69	8.99	9.16
$RMSE_u$	4.57	4.89	5.22	4.85	3.76
$RMSE_s$	7.98	6.41	5.50	7.46	7.89
$(RMSE_s/RMSE)^2$	0.73	0.60	0.51	0.69	0.74
d	0.61	0.70	0.79	0.68	0.54

temperature departures correlate negatively with yield, while July precipitation departures correlate positively with yield. These results, plus the importance of July in corn's growth cycle, suggest that July should be the focus of the investigation. Using July yields and synoptic-class frequencies, eigenvector-based synoptic typing posted the highest scores in the model comparisons. Consequently, the investigators conducted a synoptic climatology of southwestern Pennsylvania corn yield using this classification. The following is abstracted from Dilley (1992).

Eigenvector-based synoptic Type 3 relates best to yield. The average surface-pressure pattern associated with Type 3 is the Bermuda high (Figure 4.4), which reaches its peak annual frequency in July (Figure 4.5).

Compositing supports the relationship between Type 3 and corn yield (Figure 6.1). Low-, average-, and high-yield years were associated with distinct composite surface-pressure patterns in July. Below-average yields exhibited July patterns with a strong elongated high centered over the region. Above-average yields showed a weaker, more zonal pressure gradient with the center of the high displaced considerably southeast. The pressure pattern for average-yield years fell between these extremes. The configuration of the composite of all July Type 3 days is nearly identical to that for high-yield years. Thus, both compositing and the model-comparisons identified the same circulation pattern in connection with high corn yields in southwestern Pennsylvania.

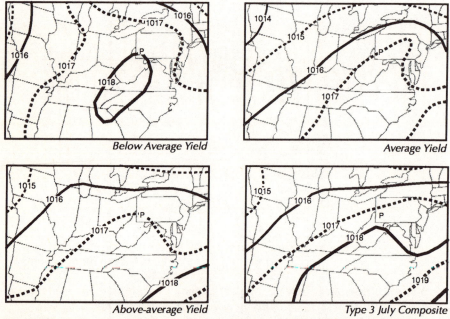

Figure 6.1 Composite maps of surface pressure (mb) associated with below-average, average, and above-average corn yields in southwestern Pennsylvania; composite map of all July days with eigenvector-based Synoptic Type 3 (modified from Dilley, 1992, with permission of Elsevier Science Publishers B.V.).

The question remains, however, as to which climatic attributes of Type 3 are beneficial to corn. The light south-southwesterly winds of a typical Type 3 day in July advect moist, warm air into the study area. These days are humid with exceptionally low visibility. Rainfall occurs on nearly 40% of Type 3 days; mean totals are 4.1 mm. During the 10-year study period, Type 3 occurred on 23% of July days. Taken together with the regression results, it appears that corn yields depend on July climate to meet the following requirements: warm, but not hot; low-to-moderate solar radiation; high precipitation; high humidity; and low wind. In other words, corn yields are highest when energy inputs are adequate but moisture stress is low. Type 3 meets these requirements, so the more Type 3 days in July, the better.

Fluvial hydrology

Humanity depends upon the fluvial system for many purposes such as drinking water, irrigation water, and sewage disposal. The flow level in rivers is critical. If it is too low, concerns about water quality, as well as quantity, move to the forefront. During periods of low discharge, greater concentrations of chemicals in stream flow can be hazardous to human health, agriculture, and aquatic ecosystems. On the other hand, floods raise the possibility of chemicals and sewage contaminating water supplies, in addition to the physical worries associated with overbank flow. Thus, water quantity and water quality are intimately linked in the fluvial system.

One of the most serious water-quality problems in many regions is related to acid rain. Sulfate concentrations in many of America's pristine streams increased substantially from the mid-1960s to the early 1980s (Lins, 1986). Observations suggest that aquatic ecosystems are stressed by these high levels of acidity.

Acidification of streams and lakes happens when acid rain alters the soil chemistry. Sulfate and nitrate anions in rainfall or snow melt provide an acidic solution to transport hydrogen and aluminum out of the soil and into the stream (Reuss and Johnson, 1985). The quantity of sulfate present is key. Small quantities of this acid will allow calcium to govern the solution, keeping the base saturation high. Increases in sulfate decrease the base saturation, allowing aluminum to dominate and causing an increase in the acidity of soil water (Reuss, 1983).

The surface waters of the eastern United States are especially prone to acidification. The soils are acidic; that is, they have a naturally low base-saturation level. The addition of any further sulfate can easily lower the base-saturation level to a point where aluminum will replace calcium as the preferred cation. In those areas where fresh waters are poorly buffered, any increase in the acid delivered to streams can overwhelm the system (Reuss *et al.*, 1987). An increase in acid rain could have a significant impact on the aquatic ecosystems of these regions.

The weather, stream level, and stream acidity appear to be linked. Several investigators have associated increased in-stream sulfate concentrations with increases in stream discharge (for example, Morris and Thomas, 1987;

Kerekes and Freedman, 1989; Seip *et al.*, 1989). Most of these studies have focused on snow melt or single-storm events (Harvey and Whelpdale, 1986; Baird *et al.*, 1987; Morris and Thomas, 1987; Abrahams *et al.*, 1989). Sulfate concentrations in surface water display considerable variation as meteorological conditions change (Lynch and Corbett, 1989). Generally, sulfate concentrations begin to rise after the start of precipitation and decline after its end.

Therefore, because the atmosphere transports acids into a region, provides the mechanism for acid deposition, and supplies the medium that leaches the acids from the soil and into the stream, understanding the weather patterns leading to a high-acid stream-flow event is important. Synoptic climatology provides methods to associate variations in the atmospheric circulation, stream levels, and stream acidity. With this scenario in mind, the investigative team decided to conduct separate synoptic climatologies of both stream flow and in-stream sulfate. They would then compare the two sets of results to suggest linkages between the fluvial variables.

The investigators settled on a data set from the Leading Ridge Experimental Watershed Research Unit in Huntingdon County, central Pennsylvania. Leading Ridge Watershed One has an areal extent of 122.6 hectares. The stream draining Leading Ridge One develops from two perennial, first-order channels and several intermittent channels. The stream exhibits a rapid response time, with increased hydrologic activity noted less than 30 minutes after precipitation starts (Lynch and Corbett, 1989).

Leading Ridge lies in the Appalachian Highlands Ridge and Valley Province, and is underlain by strongly folded sedimentary rocks. The upper slope and ridge top are underlain by quartzite, the middle slope by sandstone, and the lower slope by shale. Limestone is located in a narrow belt below the stream-gauging station. The absence of limestone above the gauging station is important, since chemical weathering of limestone could interfere with both the stream flow and water chemistry results. Most of the upland soils on Leading Ridge are residual; that is, they formed from the underlying bedrock. The natural vegetation is a second-growth hardwood forest. No known harvesting occurred during the study period that could have caused runoff levels to fluctuate. Lynch and Corbett (1989) give a complete account of the biology, geology, hydrology, and pedology for the study site.

The precipitation data are daily values collected at the site, and are continuous for the entire 10-year study period. The discharge data are daily averages recorded by the stream-gaging station; only one day is missing from this record.

There is a distinct seasonal cycle of stream discharge at Leading Ridge, with peak runoff in early spring and a low in the late summer–early autumn. However, the investigation called for a measure that is not sensitive to absolute differences in discharge. Thus, the investigators utilized the change in percentage from the previous day's discharge to capture the day-to-day variability in stream flow associated with various synoptic types. This value is also indicative of a particular synoptic event's precipitation intensity. Unfortunately, the percentage increase from the previous day can be misleading, since large changes in discharge are more likely during low-flow periods than high-flow periods. This is an admitted weakness, but all of the selected extreme events proved to be significant rain episodes.

The in-stream sulfate data are determined from weekly grab samples; that is, the specimen bottle is immersed by hand in the stream. Although these data cover the same 10-year period as the discharge data, there were some missing sulfate values. Generally, the missing data occur at the end of December and the first part of January in most years. Thus, only 479 data samples out of a possible 520 were available for analysis.

The in-stream concentration of sulfate at Leading Ridge One displays a strong seasonal cycle. The lowest stream-sulfate concentrations occur from late summer to early autumn (August and September) and increase from October to December. The highest sulfate levels are recorded from January to March and are followed by a steady decrease during the spring and summer months. This is the reverse of the sulfate concentrations in precipitation recorded at the site. The decline during the spring and summer appears to result from the decrease in base flow associated with soil-moisture depletion during the growing season. Ground-water sources control stream discharge in summer, with relatively small contributions from storm flow. This seasonal pattern suggests the existence of a sulfate reservoir in the watershed's soils (Lynch and Corbett, 1989).

A significant problem encountered by the investigative team was selecting an appropriate technique to remove the seasonal signal from the Leading Ridge in-stream sulfate observations. Since the data exhibit a periodic fluctuation, the investigators used a Fourier transformation to standardize the data. They filtered any periodic component greater then three months, thus eliminating the seasonal signal present in the data. Besides the seasonal signal, no statistically significant peaks occur in the power spectrum. Because the data set is not continuous over time, each transformed value was compared to the original sulfate data to ensure that it was not an artifact of the time-series reconstruction.

Synoptic-classification performance. The fluvial-hydrology scenario generated the worst results of the study (Tables 6.4 and 6.5). Especially bad were those associated with stream flow. Although the RMSE-based statistics look normal, d reveals huge errors in the model predictions.

Why did stream flow relate so poorly to the synoptic classifications? The answer is probably related to soil moisture. Preconditioning of the soil determines the response of a watershed to a rain event. If soil is saturated, the response is strong and reflects the magnitude of the rain event. If soil is moist, but less than field capacity, then the soil must be brought to field capacity before runoff can occur. With moist soil, runoff responds positively to large rain events, is not necessarily well matched to moderate rain events, and has little correlation with light rainfalls. If the soil is dry, only the largest rain events can bring the soil to field capacity and produce runoff. Even

Table 6.4 Synoptic-classification model performance in the fluvial-hydrology (stream-flow) scenario

	Manual synoptic types	Correlation-based map patterns	Eigenvector-based synoptic types	Eigenvector-based map patterns	Circulation index
RMSE	1.45	1.49	1.68	1.56	1.63
$RMSE_u$	0.90	0.91	0.97	0.95	0.96
$RMSE_s$	1.14	1.18	1.38	1.24	1.32
$(RMSE_s/RMSE)^2$	0.62	0.63	0.67	0.63	0.66
d	0.08	0.07	0.02	0.04	0.03

Table 6.5 Synoptic-classification model performance in the fluvial-hydrology (in-stream sulfate) scenario

	Manual synoptic types	Correlation-based map patterns	Eigenvector-based synoptic types	Eigenvector-based map patterns	Circulation index
RMSE	1.29	1.35	1.54	1.42	1.36
$RMSE_u$	0.91	0.91	0.92	0.91	0.91
$RMSE_s$	0.91	0.99	1.22	1.09	1.01
$(RMSE_s/RMSE)^2$	0.50	0.54	0.63	0.59	0.55
d	0.16	0.13	0.05	0.10	0.13

then, the association between runoff and magnitude is poor, depending on the absolute magnitude of the rain event, the storage capacity of the soil, and other factors such as infiltration rate. Thus, predictions of daily stream flow based on simple daily synoptic classes are often doomed to failure.

There are a few remedies the synoptic climatologist can use to cure this problem. First, the investigator can restrict the analysis to large-magnitude rain events. Second, he or she can test stream-flow response only when the soil is saturated. These two prescriptions, therefore, dictate switching from circulation-to-environment schemes to environment-to-circulation approaches. A third solution that permits the investigator to utilize circulation-to-environment classifications involves sequencing (Comrie, 1992b). Sequencing identifies multi-day progressions of synoptic types most strongly associated with an environmental condition. In this case, certain sequences of prolonged rainfall might always raise the soil-water content to field capacity and promote heavy runoff. Fourth, to eliminate the effect of soil on fluvial response, an investigator can search for watersheds with little soil and vegetation. In contrast, he or she can apply synoptic climatology to a watershed in a perhumid region where the soils are nearly always saturated. Both watersheds would give positive responses to inputs.

The other variable used in the fluvial-hydrology scenario, in-stream sulfate concentrations, fared better (Table 6.5). RMSE values were moderate, and unsystematic error made up a large proportion of the classification error. Moderate tuning is needed to improve model performance. Most important, d-values for sulfate concentrations, although still relatively low, are much better than those of the stream-flow results.

Relative performance of the various classifications in both the stream-flow and water-quality portions suggests that eigenvector-based synoptic typing is an inappropriate strategy for this scenario. This is not surprising because of the failure of this procedure to address precipitation and the relative unimportance of air-mass considerations to fluvial processes. At the other end of this performance spectrum, the manual synoptic types had the least prediction error, followed by correlation-based map patterns. Eigenvector-based map patterns and the synoptic index did moderately well in the water-quality scenario, but poorly in the stream-flow scenario.

Sulfate concentration in stream flow (Table 6.5) posted d-scores about half of those recorded for sulfate concentration in precipitation (Table 6.2). Given the physical and chemical mediation of the acid rain by the watershed's soil, this result makes sense.

Relationships between the atmospheric circulation and the fluvial-hydrology scenario. None of the circulation-to-environment approaches adequately explain variations in fluvial hydrology at Leading Ridge. Through an understanding of the processes involved, however, the investigative team still believed that the atmospheric circulation must be related to this environment. They concluded that a select group of events might demonstrate the association between circulation patterns and the periods of hydrologic activity needed to trigger high stream flow and high in-stream sulfate concentrations. In other words, they decided to try an environment-to-circulation approach.

To begin the environment-to-circulation classification, the investigators needed to determine appropriate criteria for classifying circulation data. As the criterion for inclusion in the high-runoff event group, they set the value as a stream-discharge increase of greater than 300% (that is, greater than 1.0 standard deviation) from the previous day. Overall, 119 days meet this criterion and were examined in further detail.

Precipitation occurred on 116 of these 119 days. A closer look at the three days on which there was no precipitation reveals that a significant amount of precipitation fell on the previous day. This suggests a problem in the timing of the precipitation observation and the recording of the synoptic chart; a problem inherent to all daily synoptic climatologies. The investigators deleted these three days from the record, effectively setting a second criterion for selection of circulation data: precipitation must be recorded. The mean precipitation total for the 116 days was 29.5 mm, with a standard deviation of 13.7 mm. The distribution of the data shows that 74 events had precipitation totals in excess of 25.4 mm (1 inch).

The manual inspection of each daily weather map associated with a high-runoff event provides many interesting results. One is that most of the events can be subdivided into two classes: migratory wave cyclones or stationary fronts. These synoptic features have distinct spatial and temporal characteristics that are important in the understanding of the Leading Ridge high stream-flow events.

The first class encompasses high-flow events caused by the nearby passage of a low-pressure system which the investigators referred to as *storm events*. These storms have well-defined warm and cold fronts, and they affect a significant portion of the eastern United States. The passage of a low-pressure center to the north can also transport large quantities of warm, moist air from the Gulf of Mexico, thus creating the potential for a large precipitation event. In addition, the northward advection of warm air has the capability to melt snow. A large melt can cause a significant increase in stream flow even if the amount of precipitation from the storm is relatively light.

The second high-discharge class is *stationary fronts* which have been present for two to five days. This category helps to explain why large increases in discharge can occur with a relatively low one-day precipitation total. In these instances, the rainfall from the previous days has saturated the soil, and much of the rain from the event day runs directly into the channel via overland flow or interflow. The temporal distribution of these two processes is clearly linked to the seasonal cycle of the circumpolar vortex. Strong midlatitude wave cyclones occur primarily during the winter and spring months, when the polar front is the dominant circulation feature in relation to Pennsylvania. During this period, the westerlies are vigorous due to the strong equator-to-pole temperature gradient, and organized wave cyclones traverse the eastern United States frequently. Thus, occurrences of high discharge should be associated with low-pressure systems in winter and spring. Stationary fronts, in contrast, happen primarily during the summer and autumn months. They occur with much greater frequency during the summer when the circumpolar vortex contracts and the polar front retreats north of Pennsylvania. Equator-to-pole temperature gradients decrease with concomitant weakening of the westerlies. Accordingly, stationary fronts are often associated with stagnant weather patterns.

Stationary fronts occur primarily during the summer and autumn, and only occasionally in the other seasons. Approximately 90% of all summer and autumn high-flow events corresponded to stationary fronts. Winter and spring experience significantly lower numbers of these features. Approximately 20% of all wintertime high-flow events were stationary fronts, while the remaining 80% were storm events.

The investigators noted the precipitation characteristics associated with storm and stationary-font categories on high-flow event days. For storms, precipitation totals were high on event days, with 23 recording precipitation totals in excess of 25.4 mm. The mean precipitation total was 27.9 mm with a standard deviation of 15.7 mm. Further analysis suggested that most of these events were rain events, with little if any snow falling. Stationary-fronts exhibit a distribution similar to the storm group. The mean precipitation level on the event day was higher and more constant than for the storm events, with an average of 29.5 mm and a standard deviation of 13.0 mm. Precipitation amounts greater than 25.4 mm occurred in 43 of the 64 stationary-front instances.

The investigative team decided to determine if there is a preferred center of cyclogenesis associated with high-discharge events. The team classified each event as storm, stationary front, or "other" episodes. Approximately 43% of the high-flow events were associated with storms, and 51% with stationary-fronts. The investigators excluded the "other" category from analysis due to its relative infrequency: seven instances over the 10-year period. Then they extracted the low-pressure centers for six consecutive days, stretching from Day $t-4$ to Day $t+1$ (relative to the event-day t), from the weather maps. The team members plotted these to determine the cyclones' source regions and paths relative to the study site.

The results show that the cyclogenetic areas most likely to cause a high-stream flow event are in the lee of the southern Rocky Mountains and over the Gulf of Mexico (Figure 6.2). Only one event was associated with a coastal low, and just one case involved an Alberta Clipper. The path traveled by the vast majority of storms takes the center of low pressure to Pennsylvania's northwest. Cyclones forming in the lee of the southern Rockies or over the Gulf tend to be in their mature stage when they reach Pennsylvania, with a great potential for high precipitation totals.

The preferred area of cyclogenesis shifts with time because of the seasonal cycle of the circumpolar vortex. In November, Colorado lows are the primary agent for high stream flows. In December and January, with the southward migration of the circumpolar vortex, the Gulf of Mexico is the predominant source of high-discharge events at Leading Ridge. After February, when the polar vortex begins to move north again, the center of action shifts back to Colorado.

Thus, for both storms and stationary fronts, high-flow events are related

Figure 6.2 Storm tracks associated with high-discharge episodes, Leading Ridge, Pennsylvania.

to the low-pressure track. If the track is to the northwest of Pennsylvania and the entrance to the jet stream is to the southwest (for example, the southern Great Plains or the Gulf of Mexico), the possibility increases for a high-flow event. The probability of high flows diminishes significantly if the track lies to the east (that is, along the coast or offshore) or the northwest (for example, the northern Great Plains or upper midwest).

In most respects, the relationships between the atmospheric circulation and sulfate concentrations at Leading Ridge are similar to the situation for stream flow. This makes sense because sulfate moves into the stream channel during hydrologic events. There are differences, however. Most important, the time of collection influences whether or not the meteorological conditions will satisfactorily explain the observed sulfate concentration. Since the majority of the readings come from the morning, sulfate levels often relate better to the previous day's weather. Thus, much of the following analysis will focus on the synoptic conditions of the day preceding the measurement (Day t-1).

To begin, the investigators divided the data into groups of high- and low-sulfate concentration events. Data with a standard deviation greater than $\pm$ 1.0 were retained for study. This resulted in 52 high-sulfate and 58 low-sulfate events. The team members further subdivided the high-sulfate group into storm and stationary-front categories. They assumed that the mixed distribution discovered in the stream-flow investigation would also be present in the sulfate data.

Manual inspection of the weather maps shows that low-sulfate events relate to a variety of weather patterns. None the less, the common thread among these events is that dry conditions prevailed for a minimum of two to three days prior to the day of sulfate measurement. Thus, hydrologic activity and sulfate delivery to the stream were at a minimum on that day.

The high-sulfate group, on the other hand, displays a distinct propensity for the occurrence of precipitation just before Day t. On Days $t-4$ and $t-3$, dry conditions generally prevailed, but precipitation fell by Days $t-2$ and $t-1$. This confirms the results of earlier studies, which concluded that the highest sulfate levels occur during periods of hydrologic activity (for example, Harvey and Whelpdale, 1986). On Days $t-4$ and $t-3$, the overall mean precipitation was 3.6 mm which resembles the average for any given day at Leading Ridge over the 10-year study period. Above-average rainfall (a mean of 7.4 mm per day) was evident on Day $t-2$. Day $t-1$ was the wettest of the five, with an average rainfall of 10.7 mm. On Day t, precipitation totals declined, with an average of 8.1 mm.

Continued stratification into stationary and storm events provides more insight into this phenomenon. Of the 52 high-sulfate events, there were 19 storms and 33 stationary fronts. The storm subset shows a striking composite sequence during a high-sulfate event (Figure 6.3). It begins with an elongated ridge over Pennsylvania that evolves into a back-of-high-pressure configuration. This sequence promotes the build-up and transport of atmospheric acids and acidifying agents, which move from the Ohio River Valley's polluted, high-emissions areas to the Leading Ridge site. As Day t approaches, the high-pressure system migrates eastward and a well-defined

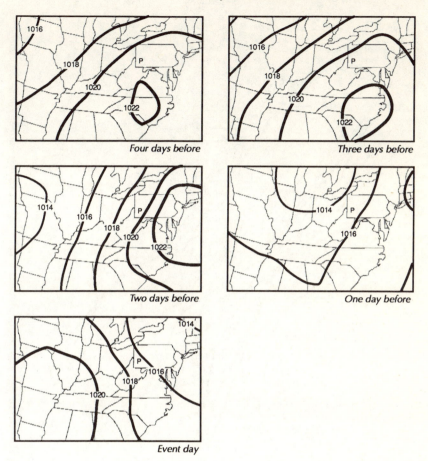

Figure 6.3 Composite maps of surface pressure (mb) associated with high-sulfate storm events, Leading Ridge, Pennsylvania.

low-pressure system from the west becomes influential. Like the earlier acid-rain scenario, this brings ample rainfall to Leading Ridge, either through decreased stability in the high's northwest quadrant or with low pressure itself. On Day t, the center of the northeastward-tracking low has passed to Pennsylvania's northwest and a high-pressure system starts to build from the southwest. This sequence of pressure patterns usually produces precipitation pH readings below 4.0 at Leading Ridge. This highly acidic precipitation releases the sulfate stored in the soil and transports it to the stream.

The investigators constructed a catalog recording the track and precipitation totals for each storm. Results demonstrate great similarity to the storm-track plots for high-stream flow events (Figure 6.2). The same general track associated with greater stream-flow levels also controls low-pressure systems related to increased in-stream sulfate concentrations (Figure 6.4). Only one coastal storm was associated with a high-sulfate

Figure 6.4 Storm tracks associated with episodes of high in-stream sulfate concentrations, Leading Ridge, Pennsylvania.

event, and most of the remainder displayed mature cyclones tracking just to the northwest of Pennsylvania.

The stationary-front events reveal a more complicated picture. On Day $t-4$, the composite pressure map displays high pressure, therefore suggesting generally dry conditions (not shown). The center of the high pressure shifts to the south on Day $t-3$, and the gentle pressure gradient suggests that the high-pressure system's influence is waning. Day $t-2$ shows the retreat of the high-pressure system out of Pennsylvania and a possible upper-level trough to the west. The approaching surface front begins to stall at this point. These conditions favor precipitation. The lack of definition in the composite pressure map suggests the muddled synoptic conditions often associated with stationary fronts. Day $t-1$ shows Pennsylvania under the influence of a trough, with a weak surface low-pressure system riding along the stationary front. The trough has moved east on Day t, and high pressure begins to reassert itself as the stationary front dissipates.

The precipitation distribution for stationary-front events differed from that of the storm events. Precipitation values reflected the prolonged, steady rainfall often associated with stationary fronts. Days $t-4$ and $t-3$ were dry a majority of the time. The effects of the approaching trough were felt on Day $t-2$, with mean rainfalls of 17.0 mm. Although Day $t-1$ experienced a greater frequency of precipitation than Day $t-2$, the average rainfall declined to 15.2 mm per day. On Day t, mean rainfall decreased to 11.2 mm.

Although the mean precipitation totals for stationary events were slightly lower than the storm averages, these events are fairly efficient sulfate producers. Their constant rainfall over a two- to three-day time span appears to cleanse the atmosphere of sulfate effectively. This is especially important during the summer months when the acid-rain concentration of atmospheric sulfate is highest. Furthermore, the constant rainfall saturates the soil and makes it easy for rainwater to transport sulfates stored in the soil to the stream channel.

Therefore certain synoptic situations are associated with increased sulfate concentrations in stream flow. A manual environment-to-circulation study suggests that the probability for such an incident increases whenever a large quantity of maritime-tropical air moves into central Pennsylvania and precipitates. For Leading Ridge, either large, powerful wintertime storm systems or summertime stationary fronts can trigger sulfate releases.

Recapitulation and discussion

The goals set out in the beginning of this chapter were (1) to evaluate the overall performance of the worked classifications, and (2) to pick which specific classification does the best job. I summarize the model-performance results in Table 6.6. The agricultural and air quality scenarios are plainly amenable to analysis by circulation-to-environment synoptic climatologies. Based on the high d-values, all of the schemes performed well in these two scenarios. The low sample standard deviations make it difficult to pick one classification over another in the air-quality scenario, while the high

Table 6.6 Summary of Willmott's d-statistic for the various environmental scenarios

	Air quality	Acid rain	Agriculture	Stream flow	In-stream sulfate
Mean d-score	0.54	0.25	0.66	0.05	0.11
Sample standard deviation of d	0.02	0.09	0.09	0.03	0.04
Best d-score	Manual synoptic types	Eigenvector-based map patte4rns	Eigenvector based synoptic types	None	None
Other good d-scores	All other classification schemes	Manual synoptic types	Correlation-based & eigenvector-based maps	None	None

standard deviations show the greater differences among schemes in the agricultural scenario. Eigenvector-based synoptic types did the best job predicting corn yields, while correlation-based and eigenvector-based map patterns made good showings, too. In the acid-rain scenario, d-scores were much lower and relatively widely dispersed. Because this scenario is based on precipitation, and precipitation is notoriously difficult to model, these numbers are not too bad. Eigenvector-based map-patterns were best here, followed by the manual types. The classifications did badly with stream flow for reasons detailed in the fluvial-hydrology section. Classification performance with water quality was also unsatisfactory, although given the difficulties suggested by the showings of the acid-rain scenario and stream flow, it did better than expected.

These results suggest a few generalizations regarding circulation-to-environment classifications. In terms of *overall performance,* this approach to synoptic climatology does credibly when dealing with problems that are holistic and do not involve mediation through daily precipitation processes. Surface ozone and crop yields appear to fit these criteria. If daily precipitation is involved, model error increases and the associations between the atmosphere and the surface environment require closer inspection. This does not mean that circulation-to-environment schemes cannot perform passably when precipitation processes are involved, but it does suggest that care must be made when interpreting the results of such a synoptic climatology. Considerable knowledge and theory must back all empirical findings and speculation. For instance, I demonstrated that acid-rain variations do relate well to variations in eigenvector-based map patterns. However, I grounded my interpretations in a literature rich in theory and observation (for instance, see Yarnal, 1991). The circulation-to-environment classifications do poorly when complex chains of variables and processes weaken the relationships between the classification and the surface environment, especially when precipitation is involved. Given the performance of numerical models in simulating complex systems and precipitation, this result is not surprising.

Regarding *relative performance*, when circulation-to-environment approaches are appropriate and can be expected to perform well, such as in the urban air-quality scenario, any classification will do an acceptable job. The superior performance of the eigenvector-based synoptic type can be rationalized in the agricultural scenario, but the small sample size indicates that this technique's fortunes might change with a larger sample, with another crop, or in another region. The important point from that scenario is that all of the circulation-to-environment classifications performed well. Thus, although proponents of various circulation-to-environment classifications tout the virtues of their methodology over all others, these may be bogus claims. The real question is whether a given environmental problem can be addressed by circulation-to-environment schemes or not. Indeed, it appears that, with fine tuning, there may be little difference among these synoptic classifications in their ability to relate the circulation and the environment. Investigators should develop methods to fine tune existing circulation-to-environment classifiications.

7   The future of synoptic climatology

Knowledge about the relationships between the atmospheric circulation and the surface environment is incomplete. One reason is that there are huge gaps in research. More synoptic climatologies of key environmental problems must be conducted. For instance, although it is uncertain how future global warming will affect crop production and water supplies, science has poor understanding of how present-day variations in weather patterns influence such basic societal needs. On the one hand, synoptic climatologists have not worked on many key environmental issues because they have not been asked to do so. On the other hand, they have not taken the initiative to identify these problems and attack them with the tools of their trade. Thus, a critical task for environmental scientists and synoptic climatologists is to work together, identifying key environmental problems and conducting collaborative research on them.

Another reason for the limited knowledge about circulation-environment relationships is that many of the difficult problems of synoptic-climatological theory and method have not been addressed adequately. I discussed synoptic-climatological theory in Chapter 1. However, throughout the book, I, like other synoptic climatologists, showed more concern about classification and applications than for the theoretical underpinnings of the field. Furthermore, I concentrated on how to use existing methods, neither showing much interest in how to improve those techniques nor developing ideas for new approaches to synoptic climatology. Nevertheless, to move synoptic climatology forward, innovative theoretical constructs and novel methodologies are needed.

In this brief concluding chapter, I address the major obstacles posed by interdisciplinary collaboration, theory building, and methodological development; I draw this framework from NRC (1992). Overcoming these barriers answers most of the theoretical questions and eliminates many of the problems faced by synoptic climatologists presented in Chapter 1.

Interdisciplinary collaboration

Synoptic climatology is an inherently multi-disciplinary, synthetic field of science. At the very least, to be a synoptic climatologist requires an investigator to have a solid grounding in atmospheric dynamics, climatic change and variability, regional climatology, and physical climatology. To be proficient at manual classification requires familiarity with synoptic meteorology; to handle automated procedures requires this, plus additional expertise in multivariate statistics and handling large, heterogeneous data sets. The investigator must keep up with advances in scale linkage. These skills enable the investigator to produce a synoptic classification and prepare him or her to relate the circulation to the surface environment.

Relating the atmospheric circulation to a complex surface environment adds another major dimension to the demands on the synoptic climatologist. To pursue a comprehensive synoptic climatology of acid rain and its environmental impacts, for instance, an investigator must understand the fields discussed above, plus the following: atmospheric chemistry, turbulent transport and diffusion, plant pathology, soil science, hydrology, aquatic biology, environmental impact assessment, and branches of social science dealing with energy, technology, and economics. A typical synoptic climatologist will try to assimilate knowledge from all these fields to carry out the research. The results will be superficial because no one aspect of the study can receive the attention it deserves.

Increasingly, environmental scientists are undertaking synoptic climatologies. Although they are steeped in the complexities of the surface environment, their knowledge of the atmosphere is limited. Stratifying the environmental data by synoptic categories improves the results, but, like the work of the lone synoptic climatologist, their findings tend to be shallow.

This one-person approach is bad for the scientist and bad for science. If the solitary investigator undertakes exhaustive synoptic-climatological research on this type of problem, then profound results cannot be produced without years of dedicated work. In the publish-or-perish environment of modern science, such single-minded attention to detail might limit the career of even the most talented scientist. In contrast, if several investigators independently pursue the same objective from different disciplinary perspectives, this results in a number of parallel, equally shallow findings. Acid-rain synoptic climatologies, for instance, have been carried out by synoptic climatologists, biologists, meteorologists, and others. All are superficial.

Therefore, for certain problems, significant, timely research can only be produced if interdisciplinary collaboration is instituted. To illustrate this type of problem, say the government wants to know what the impact would be on the environment of a region if they changed emissions standards in an effort to reduce precipitation acidity. An ideal research team would have scientists from atmospheric chemistry, turbulent transport and diffusion, synoptic climatology, plant pathology, soil science, hydrology, aquatic biology, environmental impact assessment, energy technology, fuel science, economics, and more. To simplify this scenario, assume that funds are limited and the government can only afford three investigators: a generalized environmental scientist (say, an ecologist), an atmospheric chemist specializing in transport and transformation of pollutants, and a synoptic climatologist to link the two. Such collaboration does not mean that each scientist stays strictly to his or her field. Each team member must rise to a minimum proficiency in each field, at least for that project. Thus, in the three-person acid-rain team, the synoptic climatologist must learn the rudiments of the chemistry of atmospheric acids and acidifying agents and of their production, transport and diffusion. He or she must study the basics of ecology, especially the chemical and biological processes associated with acidification of soil and water. This promotes understanding of the other scientist's research and, perhaps more important, makes it clear to the synoptic climatologist what his or her work must produce to be relevant to the others. Similarly, the ecologist must learn some synoptic climatology and atmospheric chemistry, while the chemist must study synoptic climatology and ecology for the collaboration to work. Each specialist, however, performs research in his or her field, maintaining the depth of knowledge in each area that is necessary for progress in science. Despite the period of study required of each investigator, the overall pace of research is quickened considerably and the government has its "answer" in two or three years.

As understanding of complex environmental systems grows and knowledge in each field of science deepens, such collaborative, interdisciplinary research will become more important. Indeed, it may become the norm in the policy-driven environmental research of the not-too-distant future. Synoptic climatologists are well suited for collaborative research: they are used to synthesizing information from several fields; of necessity, they have a broad knowledge base; and they are specialists in linking scales and disparate environmental variables. Because funding agencies are becoming increasingly tolerant of interdisciplinary research, now is the time for synoptic climatologists to start building bridges with other fields. It is important to note that the greatest difficulty in interdisciplinary research is getting scientists from different subject areas to talk to one another. Unfortunately, synoptic climatologists have no advantage over others in developing inter-personal relationships.

Building theory

The ultimate goal of synoptic climatology is to understand the relationships between the atmospheric circulation and the surface environment. If synoptic climatologists adopted the inductive approach to linking the atmosphere and surface, they would model the system by developing a simplified view of reality, distilling it to a number of basic equations and parameters. As their model operated, variable and parameter values would vary, but the set of variables, parameters and laws linking them would not. This assumption of invariance would keep the model from becoming too complicated to be useful.

The majority of synoptic climatologists, however, take the deductive approach to science, using statistical models to link the atmosphere and the surface environment. Typically, the investigator develops a single number, such as a map-pattern frequency or synoptic index, to reflect the atmospheric circulation for one time period. That number is then related linearly to another solitary value representing the surface environment for that span. In the case of acid rain, the investigator might use simple regression analysis to relate a synoptic index to one day's precipitation-sulfate concentrations at one site.

This is a black-box approach. It ignores all of the processes operating between the atmospheric scale of the study and the rain gauge. It omits interactions between and among physical and chemical processes; it disregards non-linearities; and it neglects linkages among the temporal and spatial scales of the myriad processes. This is, in fact, a model, but one that is simplified more than the most primitive construct of the modeler.

The synoptic climatologists of the future must change their approach from a two-variable, end-point analysis to one that incorporates a *cascade* of processes and scales (Harman and Winkler, 1991). The investigators must recognize key relationships and feedbacks among variables in that cascade; they must understand non-linear trajectories and abrupt discontinuities in physical and chemical processes, developing methods to dealing with those non-linearities; and they must formulate appropriate techniques for bridging disparate temporal and spatial scales simultaneously. Thus, synoptic climatology must move closer to the systems-oriented, stepwise approach of modeling; and synoptic climatologists must learn more atmospheric dynamics and chemistry. While such a move does not require the synoptic climatologist to abandon the deductive, empirical approach, it does require more attention to the mediating dynamic and chemical processes between the large-scale circulation and the surface environment. This, of course, places a greater load on the shoulders of the already overburdened synoptic climatologist. Because it is impossible for one investigator to know and do everything, the future of synoptic climatology is in multi-investigator, collaborative research.

Therefore, the main theoretical challenge to synoptic climatology does not require the field to abandon its scientific paradigm and goals. The assumptions (and their built-in problems) do not have to be altered significantly. Two things are required, however. First is a change in intensity. On any given synoptic-climatological problem, more work must be done, and it must be done better. The theoretical question is no longer: does the atmospheric circulation strongly influence the surface environment? The question is now: what processes enable the circulation to control the surface? Synoptic climatologists must work with other scientists to figure out the answer to this question, thus facilitating informed political decisions on actions to mitigate environmental problems and their impacts. A second requirement is new methods.

Methodological development

I devoted much of this book to current synoptic-climatological methodology. I attempted to clarify the various methods, differentiating the various approaches and schemes on theoretical, practical, and numerical bases. The model–comparison analysis of Chapter 6 showed that many of the classifications are suited to environmental analysis, but that substantial fine tuning is needed before the classification-induced error is eliminated. Removing classification-induced error isolates a pure synoptic-climatological signal. The investigator can then extract the climate signal from the time series, leaving other sources of variance for analysis, such as anthropogenic disturbance of the environment (Comrie, 1992c).

A main thrust of research in synoptic climatology must be fine tuning the existing classification schemes. For example, Comrie (1992b) found that sequences of weather types can predict surface environmental values much better than single daily types, thereby reducing the classification-induced error. In this case, the improvement to the classification model resulted from an understanding of the physical and chemical processes associated with urban air quality: pollutants build over the city and are transported into the metropolitan area when certain combinations of weather patterns occur. Thus, sequencing enhanced this particular synoptic climatology. A careful assessment of environmental and climatological processes will allow the investigator to determine such tuning mechanisms in most synoptic climatologies.

Although the traditional black-box approach should not be the main thrust of synoptic-climatological research in the future, there are still a few potentially fruitful avenues to travel. One of the most promising is that of neural networks (B.C. Hewitson, personal communication). Through an iterative numerical approach, neural networks seek and achieve the optimal statistical relationship between two suites of numbers. Applying neural networks to synoptic climatology could result in optimal synoptic classifications and the strongest possible relationships between the classes and the surface environment. The danger here is that investigators would put their efforts into producing tight circulation–surface relationships without truly understanding the processes permitting that relationship to exist.

In the long run, the only valid future for synoptic-climatological methodology is a systematic approach linking the cascade of processes and scales between the atmospheric circulation and surface environment. One possible way to achieve this goal is through geographic information systems (GIS) (Miller et al., 1991). GIS can control many variables and many spatial scales simultaneously. GIS specialists are actively working on theory and practical means to add the temporal dimension to their data matrices. Such an advance would empower the synoptic climatologist to create multidimensional classifications of time, space, and processes. By developing methods to understand and link complex non-linear systems and disparate scales, synoptic climatologists will certainly be in the mainstream, if not at the forefront, of future scientific inquiry.

References

Abler, R., Adams, J.S. and Gould, P. 1971. *Spatial Organization*. Prentice-Hall, Inc., Englewood Cliffs, New Jersey.

Abrahams, P.W., Tranter, M., Davies, T.D. and Blackwood, I.L. 1989. 'Geochemical studies in a remote Scottish upland catchment II. Streamwater chemistry during snow-melt', *Water, Air, and Soil Pollution* **43**, 231–248.

Achtor, T.H. and Horn, L.H. 1986. 'Spring season Colorado cyclones. Part I: Use of composites to relate upper and lower tropospheric wind fields', *Journal of Climate and Applied Meteorology* **25**, 732–43.

Alijani, B. and Harman, J.R. 1985. 'Synoptic climatology of precipitation in Iran', *Annals of the Association of American Geographers* **75**, 404–416.

Alt, B.T. 1978. 'Synoptic climate controls of mass-balance variations on Devon Island Ice Cap', *Arctic and Alpine Research* **10**, 61–80.

Alt, B.T. 1979. 'Investigation of summer synoptic climate controls on the mass balance of Meighen Ice Cap', *Atmosphere-Ocean* **17**, 181–199.

Alt, B.T. 1983. 'Synoptic analogs: A technique for studying climatic change in the Canadian High Arctic', in C.R. Harrington (ed.), *Climate Change in Canada 3/Syllogeus* **33**, 70–107.

Alt, B.T. 1987. 'Developing analogs for extreme mass balance conditions on Queen Elizabeth Island ice caps', *Journal of Climate and Applied Meteorology* **26**, 1605–1623.

Altshuller, A.P. 1978. 'Association of oxidant episodes with warm stagnating anticyclones', *Journal of the Air Pollution Control Association* **28**, 152–155.

Ambrozy, P., Bartholy, J. and Gulyas, O. 1984. 'A system of seasonal macrocirculation patterns for the Atlantic-European region', *Idojaras* **88**, 121–133.

Baird, S.F., Buso, D.C., and Hornbeck, J.W. 1987. 'Acid pulses from snowmelt at acidic Cone Pond, New Hampshire', *Water, Air, and Soil Pollution* **34**, 325–338.

Balling, R.C., Jr. 1984. 'Classification in climatology', in G.L. Gaile and C.M. Willmott (eds.) *Spatial Statistics and Models,* D. Reidel Publishing Company, Dordrecht, 81–108.

Barchet, W.R. 1982. *A Weather Pattern Climatology of the Great Plains and the Related Wind Regime.* Pacific Northwest Laboratory, Richland, Washington, U.S. Department of Energy, PNL-4330.

Barchet, W.R. and Davis, W.E. 1983. *Estimating Long-Term Mean Wind from Short-Term Wind Data.* Pacific Northwest Laboratory, Richland, Washington, U.S. Department of Energy, PNL-4785.

Barchet, W.R. and Davis, W.E. 1984. *A Weather Pattern Climatology of the United States.* Pacific Northwest Laboratory, Richland, Washington, U.S. Department of Energy, PNL-4889.

References

Bardossy, A. and Caspary, H.J. 1990. 'Detection of climate change in Europe by analyzing European atmospheric circulation patterns from 1881–1989', *Theoretical and Applied Climatology* **42**, 155–167.

Barry, R.G. 1980. 'Synoptic and dynamic climatology', *Progress in Physical Geography* **4**, 88–96.

Barry, R.G., Bradley, R. and Jacobs, J. 1975. 'Synoptic climatological studies of the Baffin Island area', in G. Weller and S.A. Bowling (eds.) *Climate of the Arctic,* Geophysical Institute, Fairbanks, Alaska, 339–346.

Barry, R.G., Crane, R.G., Schweiger, A. and Newell, J. 1987. 'Arctic cloudiness in spring from satellite imagery', *Journal of Climatology* **7**, 423–451.

Barry, R.G., Elliott, D.L. and Crane, R.G. 1981a. 'The palaeoclimatic interpretation of exotic pollen peaks in Holocene records from the eastern Canadian Arctic: A discussion', *Review of Palaeobotany and Palynology* **33**, 153–167.

Barry, R.G., Kiladis, G. and Bradley, R.S. 1981b. 'Synoptic climatology of the western United States in relation to climatic fluctuations during the Twentieth Century', *Journal of Climatology* **1**, 97–113.

Barry, R.G. and Perry, A.H. 1973. *Synoptic Climatology,* Methuen, London.

Blasing, T.J. 1975. 'A comparison of map-pattern correlation and principal component eigenvector methods for analyzing climatic anomaly patterns', *Fourth Conference on Probability and Statistics in Atmospheric Science,* American Meteorological Society, Boston, 96–101.

Blasing, T.J. 1979. 'Map pattern classification at a prescribed level of generality', *Sixth Conference on Probability and Statistics in Atmospheric Science,* American Meteorological Society, Boston, 118–125.

Blasing, T.J. 1981. 'Characteristic anomaly patterns of summer sea-level pressure for the Northern Hemisphere', *Tellus* **33**, 428–437.

Blasing, T.J. and Lofgren, G.R. 1980. 'Seasonal climatic anomaly types for the North Pacific sector and western North America', *Monthly Weather Review* **108**, 700–719.

Bradley, R.S. and England, J. 1979. 'Synoptic climatology of the Canadian High Arctic', *Geografiska Annaler* **61A**, 187–201.

Brazel, A.J., Kalkstein, L.S. and Chambers, F.B. 1991. 'Summer energy balance on west Gulkana Glacier, Alaska and linkages to a temporal synoptic index', *Archiv fur Geomorphologie,* in press.

Brazel, A.J. and Nickling, W.G. 1986. 'The relationship of weather types to dust storm generation in Arizona (1965–1980), *Journal of Climatology* **6**, 255–275.

Briffa, K.R., Jones, P.D. and Kelly, P.M. 1990. 'Principal component analysis of the Lamb catalogue of daily weather types: Part 2, seasonal frequencies and update to 1987', *International Journal of Climatology* **10**, 549–563.

Buell, C.E. 1975. 'The topography of empirical orthogonal functions', *Fourth Conference on Probability and Statistics in Atmospheric Science,* American Meteorological Society, Boston, 188–193.

Buell, C.E. 1979. 'On the physical interpretation of empirical orthogonal functions', *Sixth Conference on Probability and Statistics in Atmospheric Science,* American Meteorological Society, Boston, 112–117.

Carleton, A.M. 1985. 'Synoptic and satellite aspects of the southwestern U.S. summer "monsoon"', *Journal of Climatology* **5,** 389–402.

Carleton, A.M. 1986. 'Synoptic-dynamic character of "bursts" and "breaks" in the south-west U.S. summer precipitation singularity', *Journal of Climatology* **6,** 605–623.

Carleton, A.M. 1987. 'Summer circulation climate of the American Southwest, 1945–1984', *Annals of the Association of American Geographers* **77,** 619–634.

Carleton, A.M. 1992. *Satellite Remote Sensing in Climatology,* Belhaven Press, London.

Carlson, T. 1991. *Mid-Latitude Weather Systems,* Routledge, London.

Childers, D.L., Day, J.W., Jr. and Muller, R.A. 1990. 'Relating climatological forcing to coastal water levels in Louisiana estuaries and the potential importance of El Niño–Southern Oscillation events', *Climate Research* **1,** 31–42.

Christiansen, W.L. and Bryson, R.A. 1966. 'An investigation of the potential of component analysis for weather classification', *Monthly Weather Review* **94,** 697–709.

Chung, Y.S. 1978. 'The distribution of atmospheric sulphates in Canada and its relationship to long-range transport of air pollutants', *Atmospheric Environment* **12,** 1471–1480.

Comrie, A.C. 1990a. 'The climatology of surface ozone in rural areas: A conceptual model', *Progress in Physical Geography* **14,** 295–316.

Comrie, A.C. 1992a. *A Synoptic Climatology of Ozone Concentrations in the Forests of Pennsylvania.* Unpublished PhD dissertation, Pennsylvania State University.

Comrie, A.C. 1992b. 'An enhanced synoptic climatology of ozone using a sequencing technique', *Physical Geography* (in press).

Comrie, A.C. 1992c. 'A procedure for removing the synoptic climate signal from environmental data', *International Journal of Climatology* **12,** 177–183.

Comrie, A.C. and Yarnal, B. 1992. 'Relationships between synoptic-scale atmospheric circulation and ozone concentrations in metropolitan Pittsburgh, Pennsylvania', *Atmospheric Environment* (in press).

Court, A.A. 1957. 'Climatology: complex, dynamic and synoptic', *Annals of the Association of American Geographers* **47,** 125–136.

Crane, R.G. 1978. 'Seasonal variations of sea ice extent in the Davis Strait–Labrador Sea area and its relationships with synoptic-scale atmospheric circulation', *Arctic* **31,** 437–447.

Crane, R.G. 1979. 'Synoptic controls on the energy budget regimes of an ablating fast ice surface', *Archiv fur Meteorologie, Geophysik und Bioklimatologie* **28A,** 53–70.

Crane, R.G. and Barry, R.G. 1988. 'Comparison of the MSL synoptic pressure patterns of the Arctic as observed and simulated by the GISS general circulation model', *Meteorology and Atmospheric Physics* **39,** 169–183.

Cushman, R.M., Farrell, M.P. and Koomanoff, F.A. 1988. 'Climate and regional resource analysis: The effect of scale on resource homogeneity', *Climatic Change* **13**, 129–148.

Davies, T.D., Dorling, S.R., Pierce, C.E., Barthelmie, R.J. and Farmer, G. 1991. 'The meteorological control on the anthropogenic ion content of precipitation at three sites in the UK: The utility of Lamb Weather Types', *International Journal of Climatology* **11**, 795–807.

Davies, T.D., Farmer, G. and Barthelmie, R.J. 1990. 'Use of simple daily atmospheric circulation types for the interpretation of precipitation composition at a site (Eskdalemuir) in Scotland, 1978–1984', *Atmospheric Environment* **24A,** 63–72.

Davies, T.D., Kelly, P.M., Brimblecombe, P., Farmer, G. and Barthelmie, R.J. 1986. 'Acidity of Scottish rainfall influenced by climatic change', *Nature* 322, 359–361.

Davies, T.D., Palutikof, J.P., Holt T. and Kelly, P.M. 1988. 'Variations in the wind resource over the British Isles for potential power production', in S. Gregory (ed.) Recent Climatic Change — A Regional Approach, Belhaven Press, London, 69–78.

Davis, J.C. 1973. *Statistics and Data Analysis in Geology* (first edition). John Wiley and Sons, New York.

Davis, R.E. and Kalkstein, L.S. 1990a. 'Development of an automated spatial synoptic climatological classification', *International Journal of Climatology* **10**, 769–794.

Davis, R.E. and Kalkstein, L.S. 1990b. 'Using a spatial synoptic climatological classification to assess changes in atmospheric pollution concentrations', *Physical Geography* **11**, 320–342.

Dayan, U. 1986. 'Climatology of back trajectories from Israel based on synoptic analysis', *Journal of Climate and Applied Meteorology* **25,** 591–595.

Deutscher Wetterdienst 1948–1990. 'Die Großawetterlagen Europas,' *Amstblatt des Deutschen Wetterdienstes* **30,** Zentralamt, Deutscher Wetterdienst, Offenbach am Main.

Dey, B. 1982. 'Nature and possible causes of droughts on the Canadian prairies — case studies', *Journal of Climatology* **2** 233–249.

Diab, R.D., Preston-Whyte, R.A. and Washington, R. 1991. 'Distribution of rainfall by synoptic type over Natal, South Africa', *International Journal of Climatology* **11,** 877–888.

Diaz, H.F. and Namias, J. 1983. 'Associations between anomalies of temperature and precipitation in the United States and western Northern Hemisphere 700 mb height profiles', *Journal of Climate and Applied Meteorology* **22,** 352–363.

Dilley, F.B. 1992. 'The statistical relationship between weather-type frequencies and corn (maize) yields in southwestern Pennsylvania. *Agricultural and Forest Meteorology* (in press).

Douglas, A.V. 1974. *Cutoff Lows in the Southwestern United States and Their Effects on the Precipitation of this Region,* Final Report, NOAA, Contract 1-35241, Tree Ring Lab, University of Arizona, Tucson.

Douglas, A.V. and Fritts, H.C. 1972. *Tropical Cyclones of the Eastern*

North Pacific and Their Effects on the Climate of the Western U.S., Final Report, NOAA, Contract 1-35241, Tree Ring Lab, University of Arizona, Tuscon.

Englehart, P.J. and Douglas, A.V. 1985. 'A statistical analysis of precipitation frequency in the conterminous United States, including comparisons with precipitation totals', *Journal of Climate and Applied Meteorology* **24**, 350–361.

EPRI 1983. *The Sulfate Regional Experiment: Report of Findings, Volume 3.* Electric Power Research Institute, Palo Alto, California.

Ezcurra, A., Casado, H., Lacaux, J.P. and Garcia, C. 1988. 'Relationships between meteorological situations and acid rain in Spanish Basque country', *Atmospheric Environment* **22**, 2779–2786.

Fahl, C.B. 1975. 'Mean sea level pressure patterns relating to glacier activity in Alaska', in G. Weller and S.A. Bowling (eds), *Climate of the Arctic*, Geophysical Institute, Fairbanks, Alaska, 339–346.

Faiers, G.E. 1988. 'A synoptic weather type analysis of January hourly precipitation at Lake Charles, Louisiana', *Physical Geography* **9**, 223–231.

Farmer, G., Davies, T.D., Barthelmie, R.J., Kelly, P.M. and Brimblecombe, P. 1989. 'The control by atmospheric pressure patterns of sulphate concentrations in precipitation at Eskdalemuir, Scotland', *International Journal of Climatology* **9**, 181–189.

Fitzharris, B.B. 1981. 'Frequency and climatology of major avalanches at Rogers Pass, 1909 to 1977', National Research Council, Canada, Division of Building Research, *Technical Paper no. 956* (NRC 19020).

Fitzharris, B.B. and Bakkehoi, S. 1986. 'A synoptic climatology of major avalanche winters in Norway', *Journal of Climatology* **6**, 431–446.

Flocas, A.A. 1984. 'The annual and seasonal distribution of fronts over central-southern Europe and the Mediterranean', *Journal of Climatology* **4**, 255–267.

Flocas, A.A. and Giles, B.D. 1991. 'Distribution and intensity of frontal rainfall over Greece', *International Journal of Climatology* **11**, 429–442.

Forbes, G.S. and Pielke, R.A. 1985. 'Use of observational and model-derived fields and regime model output statistics in mesoscale forecasting', *ESA Journal* **9**, 207–225.

Fraedrich, K. 1990. 'European Grosswetter during the warm and cold extremes of the El Niño/Southern Oscillation', *International Journal of Climatology* **10**, 21–31.

Goldberg, A.S. 1984. *The Relationship of High Ozone Levels from Anthropogenic Sources to Synoptic Weather Types in Baton Rouge, Louisiana: 1981–1985.* Unpublished MS thesis, Louisiana State University.

Gould, P.R. 1967. 'On the geographic interpretation of eigenvalues', *Transactions of the Institute of British Geographers,* Winter, 53–86.

Gould, P.R. 1982. 'The tyranny of taxonomy', *The Sciences* **22**, 7–9.

Graham, R.L., Hunsaker, C.T., O'Neill, R.V. and Jackson, B.L. 1991. 'Ecological risk assessment at the regional scale', *Ecological Applications* **1**, 196–206.

References

Grigg, D.B. 1965. 'The logic of regional systems', *Annals of the Association of American Geographers* **55**, 465–491.

Grigg, D.B. 1967. 'Regions, models and classes', in R.J. Chorley and P. Hagget (eds.), *Models in Geography,* Methuen, London, 461–509.

Hall, F.G., Strebel, D.E. and Sellers, P.J. 1988. 'Linking knowledge among spatial and temporal scales: Vegetation, atmosphere, climate and remote sensing', *Landscape Ecology* **2**, 3–22.

Harman, J.R. 1991. 'Synoptic climatology of the westerlies: Process and patterns', *Resource Papers in Geography no. 11,* Association of American Geographers, Washington, D.C.

Harman, J.R. and Harrington, J.A. 1978. 'Contrasting rainfall patterns in the upper middle west', *Annals of the Association of American Geographers* **68**, 402–413.

Harman, J.R. and Winkler, J.A. 1991. 'Synoptic climatology: Themes, applications and prospects', *Physical Geography* **12**, 220–230.

Harnack, R.P. and Lanzante, J.R. 1985. 'Specification of United States seasonal precipitation', *Monthly Weather Review* **113**, 319–325.

Harrington, J.A. and Brown, B.J. 1985. 'A synoptic climatology of depressions in warm season precipitation profiles from the upper middle west', *Physical Geography* **5**, 186–197.

Harrington, J.A. and Harman, J.R. 1985. 'A synoptic climatology of moisture stress gradients in the western Great Lakes region', *Physical Geography* **6**, 43–56.

Harvey, D.W. 1969. *Explanation in Geography.* Edward Arnold, London.

Harvey, H.H. and Whelpdale, D.M. 1986. 'On the prediction of acid precipitation events and their effects on fishes', *Water, Air, and Soil Pollution* **30**, 579–586.

Hay, J.E. and Fitzharris, B.B. 1988. 'The synoptic climatology of ablation on a New Zealand glacier', *Journal of Climatology* **8**, 201–215.

Heathcote, J.A. and Lloyd, J.W. 1986. 'Factors affecting the isotopic composition of daily rainfall at Driby, Lincolnshire', *Journal of Climatology* **6**, 97–106.

Heidorn, K.C. and Yap, D. 1986. 'A synoptic climatology for surface ozone concentrations in southern Ontario, 1976–1981', *Atmospheric Environment* **20**, 695–703.

Hess, P. and Brezowsky, H. 1977. 'Katalog der Großwetterlagen Europas (1881–1976)', *Berichte des Deutschen Wetterdienstes,* Nr. 113 Bd. 15, Selbstverlag des Deutschen Wetterdienstes, Offenbach am Main.

Hewitson, B.C. 1992. 'Regional climate in the GISS GCM: Surface air temperature', *Journal of Climate* (submitted).

Hewitson, B.C. and Crane, R.G. 1992a. 'Regional climate in the GISS GCM: synoptic scale circulation', *Journal of Climate* (in press).

Hewitson, B.C. and Crane, R.G. 1992b. 'Regional-scale climate prediction from the GISS GCM', *Journal of Global and Planetary Change* (in press).

Hidy, G.M. 1988. 'Scientific considerations for empirical determination of regional source-receptor relationships', *Atmospheric Environment* **22**, 1801–1820.

Hirschboeck, K.K. 1987a. 'Hydroclimatically-defined mixed distributions in partial duration flood series', in V.P. Singh (ed.) *Hydrologic Frequency Modeling,* D.Reidel Publishing Company, Dordrecht, 199–212.

Hirschboeck, K.K. 1987b. 'Catastrophic flooding and atmospheric circulation anomalies', in L. Mayer and D. Nash (eds.) *Catastrophic Flooding,* Allen and Unwin, London, 23–56.

Hoard, D.E. and Lee, J.T. 1986. 'Synoptic classification of a ten-year record of 500 mb weather maps for the western United States', *Meteorology and Atmospheric Physics* **35,** 96–102.

Howarth, D.A. 1983. 'An analysis of the variability of cyclones around Antarctica and their relationship to sea-ice extent', *Annals of the Association of American Geographers* **73,** 519–537.

Jacobeit, J. 1987. 'Variations of trough positions and precipitation patterns in the Mediterranean area', *Journal of Climatology* **7,** 453–476.

Johnston, R.J. 1968. 'Choice in classification: The subjectivity of objective methods', *Annals of the Association of American Geographers* **58,** 575–589.

Jones, P.D. and Kelly, P.M. 1982. 'Principal component analysis of the Lamb Catalogue of Daily Weather Types: Part 1, annual frequencies', *Journal of Climatology* **2,** 147–157.

Kalkstein, L.S. 1979. 'A synoptic climatological approach for environmental analysis', *Proceedings of the Middle States Division of the Association of American Geographers* **13,** 68–75.

Kalkstein, L.S. 1992. 'A new approach to evaluate the impact of climate upon human mortality', *Environmental Health Perspectives* (in press).

Kalkstein, L.S. and Corrigan, P. 1986. 'A synoptic climatological approach for geographical analysis: Assessment of sulfur dioxide concentrations', *Annals of the Association of American Geographers* **76,** 381–395.

Kalkstein, L.S., Dunne, P.C. and Vose, R.S. 1990a. 'Detection of climatic change in the arctic using a synoptic climatological approach', *Journal of Climate* **3,** 1153–1167.

Kalkstein, L.S., Skindlov, J.A. and Sutherland, J.L. 1990b. 'A synoptic evaluation of winter SCENES air quality data', in C.V. Mathai (ed.) *Visibility and Fine Particles,* Transactions of the Air and Waste Management Association, 518–526.

Kalkstein, L.S., Tan, G. and Skindlov, J.A. 1987. 'An evaluation of three clustering procedures for use in synoptic climatological classification', *Journal of Climate and Applied Meteorology* **26,** 717–730.

Kalkstein, L.S. and Webber, S.R. 1990. 'A detailed evaluation of SCENES air quality data in northern Arizona using a three-dimensional synoptic approach', *Publications in Climatology XLIII,* no. 1, Laboratory of Climatology, C.W. Thornthwaite Associates/Center for Climatic Research, University of Delaware.

Karl, T.K. and Koscielny, A.J. 1982. 'Drought in the United States: 1895–1981', Journal of Climatology 2, 313–329.

Keables, M.J. 1988. 'Spatial associations of midtropospheric circulation and upper Mississippi River basin hydrology', *Annals of the Association of American Geographers* **78,** 74–92.

Keen, R.A. 1980. 'Temperature and circulation anomalies in the eastern

Canadian Arctic, summer 1946–1976', *Institute of Arctic and Alpine Research Occasional Paper no. 34,* Boulder, Colorado.

Kerekes, J. and Freedman, B. 1989. 'Seasonal variations of water chemistry in oligotrophic streams and rivers in Kejimkujik National Park, Nova Scotia', *Water, Air, and Soil Pollution* **46**, 131–144.

Key, J. and Crane, R.G. 1986. 'A comparison of synoptic classification schemes based on 'objective' procedures', *Journal of Climatology* **6**, 375–388.

Kiess, R.B. and Riordan, A.J. 1987. 'The statistical relationship between the synoptic-scale pressure field and the development and morning transition of surface inversions at two rural sites', *Journal of Climate and Applied Meteorology* **26**, 1000–1013.

Kirchhofer, W. 1973. 'Classification of European 500 mb patterns', *Arbeitsbericht der Schweizerischen Meteorologischen Zentralanstalt Nr. 43,* Geneva.

Kirchhofer, W. 1976. 'Stationsbezogene Wetterlagenklassifikation', *Veroffentlichungen der Schweizerischen Meteorologischen Zentralanstalt Nr. 34,* Geneva.

Klaus, D. and Stein, G. 1978. 'Temporal variations of the European Grosswetterlagen and possible causes', *Geophysical and Astrophysical Fluid Dynamics* **11**, 89–100.

Klein, W.H. 1983. 'Objective specification of monthly mean surface temperature from mean 700 mb heights in winter', *Monthly Weather Review* **111**, 674–691.

Klein, W.H. 1985a. 'Space and time variations in specifying monthly mean surface temperature from the 700 mb height field. *Monthly Weather Review* **113**, 277–290.

Klein, W.H. 1985b. 'Specification of monthly anomalies of surface air temperature in Canada and Alaska. *Atmosphere-Ocean,* **23** 155–176.

Klein, W.H. and Bloom, H.J. 1987. 'Specification of monthly precipitation over the United States from the surrounding 700 mb height field', *Monthly Weather Review* **115**, 2118–2132.

Klein, W.H. and Kline, J.M. 1984. 'The synoptic climatology of monthly mean surface temperature in the United States during winter relative to the surrounding 700 mb height field. *Monthly Weather Review* **112**, 433–448.

Klein, J.M. and Klein, J.M. 1986. 'Synoptic climatology of monthly mean surface temperature in the United States during summer in relation to the surrounding 700-mb height field', *Monthly Weather Review* **114**, 1231–1250.

Klein, W.H., Shabbar, A. and Yang, R. 1989. 'Specifying monthly mean surface temperatures in Canada and Alaska from the 500 mb height field', Journal of Climatology **2**, 631–638.

Klein, W.H. and Walsh, J.E. 1983. 'A comparison of pointwise screening and empirical orthogonal functions in specifying monthly surface temperature from 700 mb data', *Monthly Weather Review* **111**, 669–673.

Klein, W.H. and Yang, R. 1986. 'Specification of monthly mean surface temperature anomalies in Europe and Asia from concurrent 700 mb monthly mean height anomalies over the northern hemisphere', *Journal of Climatology* **6**, 463–484.

Knox, J.C., Bartlein, P.J., Hirschboeck, K.K. and Muckenhirn, R.J. 1975. 'The response of floods and sediment yield to climatic variation and land use in the Upper Mississippi Valley', *Institute for Environmental Studies Report 52,* University of Wisconsin, Madison.

Kozuchowski, K. and Marciniak, K. 1988. 'Variability of mean monthly temperatures and semi-annual precipitation totals in Europe in relation to hemispheric circulation patterns', *Journal of Climatology* **8**, 191–199.

Kruizinga, S. 1979. 'Objective classification of daily 500 mb patterns', *Sixth Conference on Probability and Statistics in Atmospheric Science,* American Meteorological Society, Boston, 126–129.

Ladd, J.W. and Driscol, D.M. 1980. 'A comparison of objective and subjective means of weather typing — an example from west Texas', *Journal of Applied Meteorology* **19**, 691–704.

Lamb, H.H. 1972. 'British Isles weather types and a register of the daily sequence of circulation patterns, 1861– 1971', *Geophysical Memoirs* **116**, London.

Lanicci, J.M. and Warner, T.T. 1991a. 'A synoptic climatology of the elevated mixed-layer inversion over the southern Great Plains in spring. Part I: Structure, dynamics, and seasonal evolution', *Weather and Forecasting* **6**, 181–197.

Lanicci, J.M. and Warner, T.T. 1991b. 'A synoptic climatology of the elevated mixed-layer inversion over the southern Great Plains in spring. Part II: The life cycle of the lid', *Weather and Forecasting* **6**, 198–213.

Lanicci, J.M. and Warner, T.T. 1991c. 'A synoptic climatology of the elevated mixed-layer inversion over the southern Great Plains in spring. Part III: Relationship to severe-storms climatology', *Weather and Forecasting* **6**, 214–226.

Leathers, D.J. and Palecki, M.A. 1992. 'The Pacific/North American teleconnection pattern and United States climate II: Temporal characteristics and index specification', *Journal of Climate* (in press).

Leathers, D.J., Yarnal, B. and Palecki, M.A. 1991. 'The Pacific/North American teleconnection pattern and United States climate I: Regional correlations', *Journal of Climate* **4**, 517–528.

Lindsey, C.G. 1980. *Analysis of coastal wind energy regimes.* Unpublished M.Sc. thesis, Department of Environmental Sciences, University of Virginia.

Lins, H.F. 1986. 'Recent patterns of sulfate variability in pristine streams', *Atmospheric Environment* **20**, 367–375.

Lorenz, E.N. 1956. *Empirical Orthogonal Functions and Statistical Weather Prediction,* Scientific Report No. 1, Contract AF19 (604)–1566, Meteorology Department, Massachusetts Institute of Technology, Cambridge, MA.

Lowry, W.P. and Probald, F. 1978. 'An attempt to detect the effects of steelworks on precipitation amounts in central Hungary', *Journal of Applied Meteorology* **17**, 964–975.

Lund, I.A. 1963. 'Map-pattern classification by statistical methods', *Journal of Applied Meteorology 2,* 56–65.

Lydolph, P.E. 1957. 'How many climatologies are there?', *Professional Geographer* **9,** 5–7.

Lynch, J.A. and Corbett, E.S. 1989. 'Hydrologic control of sulfate mobility in a forested watershed', *Water Resources Research* **25,** 1695–1703.

Maheras, P. 1984. 'Weather-type classification by factor analysis in the Thessaloniki area', *Journal of Climatology* **4,** 437–443.

Maheras, P. 1988. 'The synoptic weather types and objective delimitation of the winter period in Greece', *Weather* **43,** 40–45.

Maheras, P. 1989. 'Delimitation of the summer-dry period in Greece according to the frequency of weather-types', *Theoretical and Applied Climatology* **39,** 171–176.

Mass, C.F. 1991. 'Synoptic frontal analysis: Time for a reassessment?', *Bulletin of the American Meteorological Society* **72,** 348–363.

The MAP3S/RAINE Research Community 1982: 'The MAP3S/RAINE precipitation chemistry network: Statistical overview for the period 1976–1980', *Atmospheric Environment* **16,** 1603–1621.

McCabe, G.J., Jr., Hay, L.E., Kalkstein, L.S., Ayers, M.A. and Wolock, D.M. 1989. 'Simulation of precipitation by weather-type analysis', *Hydraulic Engineering '89, Proceedings of the National Conference on Hydraulic Engineering, New Orleans, Louisiana, August 14–18,* 679–684.

McCabe, G.J., Jr. and Muller, R.A. 1987. 'Synoptic weather types: An index of evaporation in southern Louisiana', *Physical Geography* **8,** 99–112.

McCutchan, M.H. 1978. 'A model for predicting synoptic weather types based on model output statistics', *Journal of Applied Meteorology* **7,** 1466–1475.

McCutchan, M.H. 1980. 'Use of cluster analysis and discriminant analysis to classify synoptic weather types in southern California', *Eighth Conference on Weather Forecasting and Analysis,* American Meteorological Society, Boston, 310–315.

McCutchan, M.H. and Schroder, M.J. 1973. 'Classification of meteorological patterns in southern California by discriminant analysis', *Journal of Applied Meteorology* **12,** 571–577.

Meentemeyer, V. 1989. 'Geographical perspectives of space, time, and scale', *Landscape Ecology* **3,** 163–173.

Michaels, P.J. 1985. 'Anomalous mid-atmospheric heights and persistent thunderstorm patterns over Florida', *Journal of Climatology* **5,** 529–542.

Miller, D.A., Yarnal, B. and Petersen, G.W. 1991. 'GIS as an integrative tool in Earth System Science research', *Preprints of the Seventh International Conference on Interactive Information and Processing Systems in Meteorology, Oceanography and Hydrology, New Orleans, Louisiana, January 1991,* Boston, American Meteorological Society, 273–275.

Miron, O. and Lindesay, J.A. 1983. 'A note on changes in airflow patterns between wet and dry spells over South Africa, 1963 to 1979', *South African Geographical Journal* **65,** 141–148.

Moldan, B., Kopacek, J. and Kopacek, J. 1988. 'Chemical composition of atmospheric precipitation in Czechoslovakia, 1978–1984 — II. Event samples', *Atmospheric Environment* **22,** 1901–1908.

Moritz, R.E. 1979. 'Synoptic climatology of the Beaufort Sea coast', *Institute of Arctic and Alpine Research Occasional Paper no. 30,* Boulder, Colorado.

Morris, E.M. and Thomas, A.G. 1987. 'Transient acid surges in an upland stream', *Water, Air, and Soil Pollution* **34,** 429–438.

Moses, T., Kiladis, G.N., Diaz, H.F. and Barry, R.G. 1987. 'Characteristics and frequency of reversals in mean sea level pressure in the North Atlantic sector and their relationship to long-term temperature trends', *Journal of Climatology* **7,** 13–30.

Muller, R.A. 1977. 'A synoptic climatology for environmental baseline analysis: New Orleans', *Journal of Applied Meteorology* **16,** 20–33.

Muller, R.A. and Faiers, G.E. 1984. *A climatic perspective of Louisiana floods during 1982–1983.* Geoscience Publications, Department of Geography and Anthropology, Louisiana State University, Baton Rouge.

Muller, R.A. and Jackson, A.L. 1985. 'Estimates of climatic air quality potential at Shreveport, Louisiana', *Journal of Climate and Applied Meteorology* **24,** 293–301.

Muller, R.A. and Tucker, N.L. 1986. 'Climatic opportunities for the long-range migration of moths', in A.N. Sparks (ed.) *Long-range migration of moths of agronomic importance to the United States and Canada: Specific examples of occurrence and synoptic weather patterns conducive to migration,* United States Department of Agriculture, Agricultural Research Service, ARS-43, 61–83.

Muller, R.A. and Wax, C.L. 1977. 'A comparative synoptic climatic baseline for coastal Louisiana', *Geoscience and Man* **18,** 121–129.

Muller, R.A. and Willis, J.E. 1983. 'New Orleans weather 1961–1980: A climatology by means of synoptic weather types', *Miscellaneous Publication 83-1,* School of Geoscience, Louisiana State University, Baton Rouge.

NRC 1992. *Global Environmental Change: Understanding the Human Dimensions,* P.C. Stern, O.R. Young and D. Druckman (eds.), National Research Council, National Academy Press, Washington, D.C.

Nicholsen, S.E. 1979. 'Statistical typing of rainfall anomalies in subSaharan Africa. *Erdkunde* **33,** 95–103.

NOAA (1978–1987) *Daily Weather Maps, Weekly Series.* National Oceanic and Atmospheric Administration, Washington D.C.

O'Neill, R.V., Johnson, A.R. and King, A.W. 1989. 'A hierarchical framework for the analysis of scale', *Landscape Ecology* **3,** 193–205.

Overland, J.E. and Hiester, T.R. 1980. 'Development of a synoptic climatology for the northeast Gulf of Alaska', *Journal of Applied Meteorology* **19,** 1–14.

Overland, J.E. and Pease, C.H. 1982. 'Cyclone climatology of the Bering Sea and its relation to sea ice extent', *Monthly Weather Review* **110,** 5–13.

Palutikof, J.P., Kelly, P.M., Davies, T.D and Halliday, J.A. 1987. 'Impacts of spatial and temporal windspeed variability on wind energy output',

Journal of Climate and Applied Meteorology **26,** 1124–1133.

Perry, A.H. 1983. 'Growth points in synoptic climatology', *Progress in Physical Geography* **7,** 90–96.

Petzold, D.E. 1982. 'The summer weather types of Quebec–Labrador', *McGill Subarctic Research Paper no. 34, Climatological Research Series no. 14,* McGill University, Montreal.

Philander, S.G.H. 1990. *El Niño, La Niña and the Southern Oscillation.* Academic Press, New York.

Pielke, R.A. 1982. 'The role of mesoscale numerical models in very-short-range forecasting', in K. Browning (ed.), *Nowcasting,* Academic Press, New York, 207–221.

Preisendorfer, R.W. 1988. *Principal Component Analysis in Meteorology and Oceanography,* posthumously compiled and edited by C.D. Mobley. Elsevier, New York.

Prezerakos, N.G. 1985. 'The northwest African depressions affecting the south Balkans', *Journal of Climatology* **5,** 643–654.

Prezerakos, N.G. and Angouridakis, V.E. 1984. 'Synoptic consideration of snowfall in Athens', *Journal of Climatology* **4,** 269–285.

Rayner, J.N. 1984. 'Simulation models in climatology', in G.L. Gaile and C.J. Willmott (eds.), *Spatial Statistics and Models,* D. Reidel Publishing Company, Dordrecht, 417–442.

Raynor, G.S. and Hayes, J.V. 1981. 'Acidity and conductivity of precipitation on central Long Island, New York in relation to meteorological variables', *Water, Air and Soil Pollution* **15,** 229–245.

Raynor, G.S. and Hayes, J.V. 1982. 'Variation in chemical wet deposition with meteorological conditions', *Atmospheric Environment* **16,** 1647–1656.

Reuss, J.O. 1983. 'Implications of the calcium–aluminum exchange system for the effects of acid precipitation on soils', *Journal of Environmental Quality* **12,** 591–595.

Reuss, J.O., Cosby, B.J., and Wright, R.F. 1987. 'Chemical processes governing soil and water acidification', *Nature* **329,** 27–32.

Reuss, J.O. and Johnson, D.W. 1985. 'Effects of soil processes on the acidification of water by acid deposition', *Journal of Environmental Quality* **14,** 26–31.

Richman, M.B. 1981. 'Obliquely rotated principal components — an improved meteorological map typing technique?' *Journal of Applied Meteorology* **20,** 1145–1159.

Richman, M.B. 1986. 'Rotation of principal components', *Journal of Climatology* **6,** 293–335.

Robinson, E. and Boyle, R.J. 1979. 'Synoptic weather typing and its application to air quality in the St. Louis Missouri area', *Proceedings of the Seventy-Second Annual Meeting of the Air Pollution Control Association, Cincinnati, Ohio, June 24–29.*

Rogers, J.C. 1983. 'Spatial variability of Antarctic temperature anomalies and their association with the Southern Hemisphere atmospheric circulation', *Annals of the Association of American Geographers* **73,** 502–518.

Rosen, R. 1989. 'Similitude, similarity, and scale', *Landscape Ecology* **3,** 207–216.

Rosswall, T., Woodmansee, R.G. and Risser, P.G. 1988. *Scales and Global Change.* John Wiley and Sons, London.

Sabin, R.C. 1974. 'Computer map typing — optimizing the correlation coefficient threshold', Air Weather Service (MAC), United States Air Force, *Technical Paper* 74–2.

Sanchez, M.L., Pascual, D., Ramos, C. and Perez, I. 1990. 'Forecasting particulate pollutant concentrations in a city from meteorological variables and regional weather patterns', *Atmospheric Environment* **24A,** 1509–1519.

SAS Institute, Inc. 1985. *SAS User's Guide: Statistics, Version 5 Edition.* SAS Institute, Inc., Cary, North Carolina.

Scholefield, P.R. 1973. 'Comparing correlations between weather maps with similar isobaric configurations but varying pressure intensities', Air Weather Service (MAC), United States Air Force, *Technical Paper 73–8.*

Schwartz, M.D. 1991. An integrated approach to air mass classification in the north central United States', *Professional Geographer* **43,** 77–91.

Seip, H.M., Andersen, D.O., Christophersen, N., Sullivan, T.J., and Vogt, R.D. 1989. 'Variations in concentrations of aqueous aluminum and other chemicals species during hydrological episodes at Birkenes, southernmost Norway', *Journal of Hydrology* **108,** 387–405.

Semple, R.K. and Green, M.B. 1984. 'Classification in human geography', in G.L. Gaile and C.J. Willmott (eds.), *Spatial Statistics and Models,* D. Reidel Publishing Company, Dordrecht, 55–79.

Sharon, D. and Ronberg, B. 1988. 'Intra-annual weather fluctuations during the rainy season in Israel', in S. Gregory (ed.) *Recent Climatic Change — A Regional Approach,* Belhaven Press, London, 102–115.

Singh, B., Nobert, M. and Zwack, P. 1987. 'Rainfall acidity as related to meteorological parameters in northern Quebec', *Atmospheric Environment* **22,** 825–842.

Singh, S.V., Mooley, D.A. and Kripalani, R.H. 1978. 'Synoptic climatology of the daily 700 mb summer monsoon flow patterns over India', *Monthly Weather Review* **106,** 510–525.

Skeeter, B.R. and Parker, A.J. 1985. 'Synoptic control of regional temperature trends in the coterminous United States between 1949 and 1981', *Physical Geography* **6,** 69–84.

Smithson, P.A. 1986. 'Synoptic and dynamic climatology', *Progress in Physical Geography* **10,** 100–110.

Smithson, P.A. 1987. 'Developments in synoptic and dynamic climatology', *Progress in Physical Geography* **11,** 121–132.

Smithson, P.A. 1988. 'Dynamic and synoptic climatology', *Progress in Physical Geography* **12,** 119–129.

Sowden, I.P. and Parker, D.E. 1981. 'A study of climatic variability of daily central England temperatures in relation to the Lamb synoptic types', *Journal of Climatology* **1,** 3–10.

Stenning, A.J., Banfield, C.E. and Young, G.J. 1981. 'Synoptic controls over katabatic layer characteristics above a melting glacier', *Journal of*

Climatology **1,** 309–324.

Stone, R.C. 1989. 'Weather types at Brisbane, Queensland: An example of the use of principal components and cluster analysis', *International Journal of Climatology* **9,** 3–32.

Suckling, P.W. and Hay, J.E. 1978. 'On the use of synoptic weather map typing to define solar radiation regimes', *Monthly Weather Review* **106,** 1521–1531.

Sumner, G. and Bonell, M. 1986. 'Circulation and daily rainfall in the north Queensland wet seasons 1970–1982', *Journal of Climatology* **6,** 531–549.

Suter, G.W. II, 1990. 'Endpoints for regional ecological risk assessments', *Environmental Management* **14,** 9–23.

Todhunter, P.E. 1989. 'An approach to the variability of urban surface energy budgets under stratified synoptic weather types', *International Journal of Climatology* **9,** 191–201.

Tsui, H.T. and Lam, K.P. 1979. 'On the problems of classification of weather contour maps', *Proceedings of the Fourth International Joint Conference on Pattern Recognition,* Institute of Electrical and Electronics Engineers, Kyoto, Japan, 635–637.

Turner, M.G., Dale, V.H. and Gardner, R.H. 1989. 'Predicting across scales: Theory development and testing', *Landscape Ecology* **3,** 245–252.

Tyson, P.D. 1981. 'Atmospheric circulation variations and the occurrence of extended wet and dry spells over southern Africa', *Journal of Climatology* **1,** 115–130.

Tyson, P.D. 1984. 'The atmospheric modulation of extended wet and dry spells over South Africa, 1958–1978', *Journal of Climatology* **4,** 621–635.

Tyson, P.D. 1986. *Climatic Change and Variability in Southern Africa.* Oxford University Press, Oxford.

Tyson, P.D. 1988. 'Synoptic circulation types and climatic variation over southern Africa', in S. Gregory (ed.) *Recent Climatic Change — A Regional Approach,* London, Belhaven Press, 202–214.

Van Dijk, W. and Jonker, P.J. 1985. 'Statistical remarks on European weather types', *Archives for Meteorology, Geophysics and Bioclimatology* **35B,** 277–306.

Vukovich, F.M. and Fishman, J. 1986. 'The climatology of summertime O_3 and SO_2 (1977–1981)', *Atmospheric Environment* **20,** 2423–2433.

Walsh, J.E., Tucek, D.R. and Richman, M.B. 1982. 'Seasonal snow cover and short-term climatic fluctuations over the United States', *Monthly Weather Review* **110,** 1474–1485.

Wax, C.L., Borengasser, M.J. and Muller, R.A. 1978. *Barataria Basin: Synoptic weather types and environmental responses.* Coastal Zone Management Series, Center for Wetland Resources, Louisiana State University, Baton Rouge.

Westerman, G.S. and Oliver, J.E. 1985. 'Upper air patterns and winter conditions in the Midwest', *Professional Paper No. 17,* Department of Geography and Geology, Indiana State University, Terra Haute, 3–21.

Whetton, P.H. 1988. 'A synoptic climatological analysis of rainfall variability in south-eastern Australia', *Journal of Climatology* **8,** 155–177.

White, D., Richman, M. and Yarnal, B. 1991. 'Climate regionalization and rotation of principal components', *International Journal of Climatology* **11,** 1–25.

Wigley, T.M.L. and Jones, P.D. 1987. 'England and Wales precipitation: A discussion of recent changes in variability and an update to 1985', *Journal of Climatology* **7,** 231–246.

Willmott, C.J. 1981. 'On the validation of models', *Physical Geography* **2,** 184–194.

Willmott, C.J. 1982. 'Some comments on the evaluation of model performance', *Bulletin of the American Meteorological Society* **63,** 1309–1313.

Willmott, C.J. 1984. 'On the evaluation of model performance in physical geography', in G.L. Gaile and C.J. Willmott (eds.), *Spatial Statistics and Models,* D. Reidel Publishing Company, Dordrecht, 443–460.

Willmott, C.J. 1987. 'Synoptic weather-map classification: Correlation versus sums-of-squares', *Professional Geographer* **39,** 205–207.

Willmott, C.J., Ackleson, S.G., Davis, R.E., Feddema, J.J., Klink, K.M., Legates, D.R., O'Donnell, J. and Rowe, C.M. 1985. 'Statistics for the evaluation and comparison of models', *Journal of Geophysical Research* **90,** 8995–9005.

Winkler, J.A. 1988. 'Climatological characteristics of summertime extreme rainstorms in Minnesota', *Annals of the Association of American Geographers* **78,** 57–73.

Yarnal, B. 1984a. 'The effect of weather map scale on the results of a synoptic climatology', *Journal of Climatology* **4,** 481–493.

Yarnal, B. 1984b. 'A procedure for the classification of synoptic weather maps from gridded atmospheric pressure surface data', *Computers and Geosciences* **10,** 397–410.

Yarnal, B. 1984c. 'Synoptic-scale atmospheric circulation over British Columbia in relation to the mass balance of Sentinel Glacier', *Annals of the Association of American Geographers* **74,** 375–392.

Yarnal, B. 1984d. 'Relationships between synoptic-scale atmospheric circulation and glacier mass balance in south-western Canada during the International Hydrological Decade, 1965–74', *Journal of Glaciology* **30,** 188–198.

Yarnal, B. 1985a. 'A 500mb synoptic climatology of Pacific north-west coast winters in relation to climatic variability, 1948–49 to 1977–78', *Journal of Climatology* **5,** 237–252.

Yarnal, B. 1985b. 'Extratropical teleconnections with El Niño/Southern Oscillation (ENSO) events', *Progress in Physical Geography* **9,** 315–352.

Yarnal, B. 1989. 'Climate'. In D.J. Cuff, W.J. Young, E.K. Muller, W. Zelinsky and R.F. Abler (eds.), *The Atlas of Pennsylvania,* Philadelphia, Temple University Press, 26–30.

Yarnal, B. 1991. 'The climatology of acid rain'. In S.K. Majumdar, E.W. Miller and J. Cahir (eds.), *Air Pollution: Environmental Issues and Health Effects,* Pennsylvania Academy of Science, 155–169.

Yarnal, B., Crane, R.G., Carleton, A.M. and Kalkstein, L.S. 1987. 'A new challenge for climate studies in geography', *Professional Geographer* **39,** 465–473.

References

Yarnal, B. and Diaz, H.F. 1986. 'Relationships between extremes of the Southern Oscillation and the winter climate of the Anglo-American Pacific coast', *Journal of Climatology* **6,** 197–219.

Yarnal, B. and Henderson, K.G. 1989. 'A climatology of polar low cyclogenetic regions in the North Pacific Ocean', *Journal of Climate* **2,** 1476–1491.

Yarnal, B. and Kiladis, G. 1985. 'Tropical teleconnections associated with El Niño/Southern Oscillation (ENSO) events', *Progress in Physical Geography* **9,** 524–558.

Yarnal, B. and Leathers, D.J. 1988: 'Relationships between interdecadal and interannual climatic variations and their effect on Pennsylvania climate', *Annals of the Association of American Geographers* **78,** 624–641.

Yarnal, B. and White, D.A. 1987. 'Subjectivity in a computer-assisted synoptic climatology I: Classification results', *Journal of Climatology* **7,** 119–128.

Yarnal, B., White, D.A. and Leathers, D.J. 1988. 'Subjectivity in a computer-assisted synoptic climatology II: Relationships to surface climate', *Journal of Climatology* **8,** 227–239.

Yu, C.-H. and Pielke, R.A. 1986. 'Mesoscale air quality under stagnant synoptic cold season conditions in the Lake Powell area', *Atmospheric Environment* **20,** 1751–1762.

Zangvil, A. and Druian, P. 1990. 'Upper air trough axis orientation and the spatial distribution of rainfall over Israel', *International Journal of Climatology* **10,** 57–62.

Index